Efficient Intelligence with Applications in Embedded Sensing

Efficient Intelligence with Applications in Embedded Sensing

Editors

Yong Liu
Xingxing Zuo

MDPI • Basel • Beijing • Wuhan • Barcelona • Belgrade • Manchester • Tokyo • Cluj • Tianjin

Editors
Yong Liu
Zhejiang University
Hangzhou, China

Xingxing Zuo
Technical University of
Munich
Munich, Germany

Editorial Office
MDPI
St. Alban-Anlage 66
4052 Basel, Switzerland

This is a reprint of articles from the Special Issue published online in the open access journal *Sensors* (ISSN 1424-8220) (available at: https://www.mdpi.com/journal/sensors/special_issues/efficient_embedded_sensing).

For citation purposes, cite each article independently as indicated on the article page online and as indicated below:

LastName, A.A.; LastName, B.B.; LastName, C.C. Article Title. *Journal Name* **Year**, *Volume Number*, Page Range.

ISBN 978-3-0365-7986-3 (Hbk)
ISBN 978-3-0365-7987-0 (PDF)

Contents

About the Editors

Yong Liu

Professor Yong Liu received the B.S. degree in computer science and engineering, and the Ph.D. degree in computer science from Zhejiang University, Hangzhou, China, in 2001 and 2007, respectively. He is currently a Professor with the Institute of Cyber-Systems and Control, College of Control Science and Engineering, Zhejiang University. His research interests include machine learning, robotics vision, multiple-sensor fusion, and intelligent systems.

Xingxing Zuo

Xingxing Zuo received the B.Eng. degree in mechanical engineering from the University of Electronic Science and Technology of China (UESTC), Chengdu, China, in 2016, and the Ph.D. degree in control science and engineering from Zhejiang University, Hangzhou, China, in 2021. He is currently a Postdoctoral Researcher with the Technical University of Munich, Munich, Germany. His research interests include computer vision, state estimation, sensor fusion, deep learning, localization, and mapping for autonomous robots in complex environments. Dr. Zuo was the recipient of the ICRA 2021 Best Paper Award in Robot Vision (Finalist).

sensors

Editorial

Efficient Intelligence with Applications in Embedded Sensing

Xingxing Zuo [1] and Yong Liu [2,*]

1   School of Computation, Information and Technology, Technical University of Munich, 85748 Garching, Germany; xingxing.zuo@tum.de
2   Institute of Cyber-Systems and Control, Zhejiang University, Hangzhou 310027, China
*   Correspondence: yongliu@iipc.zju.edu.cn

1. Introduction

Despite the fact that computational technology continues to rapidly develop, edge devices and embedded systems are still limited in terms of their computation resources due to such factors as power consumption, physical size constraints, and manufacturing cost. This poses a challenge for critical applications such as mobile robots, cell phones, and AR and VR devices, which require efficient sensing with sensors and on-board computational resources. To effectively process the abundance of sensor measurements using resource-constrained computation platforms, there is a need to limit the computation complexity of the methods deployed. This is true whether the method is data-driven or principle-driven, and high efficiency is typically a critical requirement.

This Special Issue is focused on both practical and theoretical technologies in the field of efficient intelligence and how they can be applied to diverse embedded devices such as industrial robots, unmanned vehicles, and fuel cells. The ten research papers published in this Special Issue cover a wide range of topics, including collaborative autonomous navigation with unmanned surface and aerial vehicles, multi-modal simultaneous localization and mapping (SLAM), target object tracking, LiDAR point cloud loop closure detection, motion distortion compensation for LiDAR point cloud, hybrid prognostic methods for proton-exchange-membrane fuel cells (PEMFC), detection of fabric defects during factory manufacturing, state recognition of elevator traction machines, efficient object detection neural networks, accurate pantograph detection for high-speed railways, and vision-based autonomous forklifts. It is our hope that these published papers will be beneficial for both academic researchers and relevant industrial practitioners alike.

2. Overview of Contribution

To ensure that an unmanned surface vehicle (USV) can navigate safely in complex scenarios with many obstacles, Huang et al. [1] proposed a system that involves multiple agents collaborating together. This system includes an unmanned aerial vehicle (UAV) that acts as a perceptive agent with a large receptive field, allowing it to detect obstacles from above and inform the USV of their locations. Next, a graph search-based hybrid A* planning algorithm generates an obstacle-free trajectory for the USV. This initial trajectory is further optimized by taking into account the dynamic constraints of the underactuated USV. By doing so, the planned trajectory is tailored to the USV's dynamics, making it easier for the vehicle to follow. Finally, a nonlinear model predictive controller (NMPC) with the lowest energy consumption constraint is used to control the USV and ensure it follows the planned trajectory precisely. The effectiveness and efficiency of this collaborative system have been demonstrated in a simulated environment.

Chen et al. [2] proposed a heterogenous Simultaneous Localization and Mapping (SLAM) system that combines sensor measurements from LiDAR, cameras, Inertial Measurement Units (IMU), and Global Positioning Systems (GPS). This system has three component state estimation subsystems: LiDAR-inertial odometry, visual-inertial odometry, and GPS-inertial odometry. The navigation states estimated from these subsystems

Citation: Zuo, X.; Liu, Y. Efficient Intelligence with Applications in Embedded Sensing. *Sensors* **2023**, *23*, 4816. https://doi.org/10.3390/s23104816

Received: 4 May 2023
Accepted: 8 May 2023
Published: 17 May 2023

are then fused in a pose graph optimization. This heterogenous hybrid SLAM system is designed to provide accurate and robust pose estimations, even in complex environments and difficult situations, such as when there are component sensor failures or intermittent GPS measurements. Additionally, based on the estimated camera poses from the SLAM system, an object tracking and localization module has been developed. This object tracking and localization module utilizes YOLOv4 to detect objects of interest from images captured by the camera. The detected objects are then tracked across multiple images using L-K optical flow and a Kalman filter. With the known 6-DoF relative transformation between the camera and LiDAR, the depth of the object can be obtained by the projected LiDAR points. This allows for the retrieval of 3D locations of the tracked objects. Experimental results in real-world scenarios have demonstrated the accurate pose estimation of the sensor rig, as well as the feasibility of tracking object locations using the presented multi-modal system.

When using odometry systems, there will inevitably be a slight drift in the estimation of the robot's position. This is because minor errors in the relative poses between consecutive frames accumulate over time, leading to significant deviations in the long run. To fix this problem, loop closures can be used. When the robot reaches a previously visited position, a loop closure is triggered, and the accumulated drift during that loop can be corrected. Tian et al. [3] proposed a method to improve loop closure detection in 3D LiDAR scans by using an object segmentation technique. The method uses a Scan Context descriptor, which is a global descriptor that records statistics of the 3D structure captured by LiDAR and stores the descriptors in an indexed KD-Tree. Loop closure candidates are identified as scans with small descriptor distance to existing scans. To enhance the performance of loop closure detection in complex environments, the method uses object segmentation to remove disturbances caused by unstructured objects such as cluttered vegetation. Experimental results on the KITTI dataset demonstrate that the proposed method outperforms the other compared methods.

To provide a 360-degree panoramic perception of the environment, mechanical 3D LiDARs use spinning laser sensors. However, if the LiDAR is moving during the spinning of the laser sensors, motion distortion can occur in the LiDAR scan. To solve this problem, Wu et al. [4] proposed a method that fuses IMU and wheel odometer measurements to compensate for the motion distortion in LiDAR scans. The positional displacement from the wheel odometer measurements and rotation changes from the IMU measurements are combined to estimate the 6 degrees of freedom (DoF) pose of the LiDAR. To roughly remove the motion distortion of a LiDAR scan, the pose estimations are linearly interpolated to obtain the interpolated poses for individual LiDAR points. Then, the roughly undistorted LiDAR scans are registered with each other via ICP (Iterative Closest Point), and the relative poses between the LiDAR scans are calculated. The relative poses obtained from the registration process are used to further undistort the motion distortion in LiDAR scans. Extensive experiments have shown that this proposed method is effective and feasible in compensating for motion distortion in LiDAR scans.

Xia et al. [5] proposed a new method for predicting the long-term voltage degradation of proton-exchange-membrane fuel cells (PEMFC) using a hybrid prognostic approach. The voltage measured from the PEMFC is decomposed into two components: the calendar aging component and the reversible aging component. To predict the overall aging trend of the PEMFC based on the calendar aging component, an adaptive extended Kalman filter is used. Additionally, a Long Short-Term Memory (LSTM) neural network is utilized to predict both voltage components together. The combination of the Kalman filter and LSTM helps to accurately predict the long-term voltage degradation of PEMFC. Furthermore, to improve the accuracy of the forecast, a dedicated three-dimensional aging factor is introduced into the physical aging model. Experimental results show that the proposed hybrid prognostic method delivers accurate long-term voltage-degradation prediction results, demonstrating its effectiveness over other methods.

Detecting fabric defects during factory manufacturing is crucial for ensuring high-quality products. Lin et al. [6] proposed an intelligent and efficient method for detecting fabric defects based on the YOLOv5 neural network. To overcome the challenges of detecting small and unbalanced defect patches, they modified the baseline YOLOv5 network using the Swin transformer backbone. They also incorporated a sliding-window multi-head self-attention mechanism to enhance accuracy in addition to the convolutional neural network. Furthermore, to improve detection accuracy, even on small defects, they introduced a detection layer capable of detecting 4×4 small targets, enabling detection at four different scales. To address the issue of imbalanced training samples, they used a generalized focal loss to help the model learn from positive samples. The proposed neural network was rigorously tested through ablation studies to analyze the effectiveness of each introduced network component. The experimental results demonstrate the high detection accuracy and real-time capability of the proposed neural network, making it a useful tool for fabric defect detection in factory manufacturing.

The condition monitoring and fault diagnosis of elevator traction machines is incredibly important for ensuring the safety of elevator users. Li et al. [7] proposed a new method for recognizing the state of a traction machine based on analyzing its vibration signals. To do this, they use a novel demodulation method that involves time-frequency analysis and principal component analysis. In order to extract the important modulation characteristics of the traction machine vibration signal, which can be difficult due to background noise interference, the researchers employ two methods: Fast Fourier Transform (FFT) and Short-Time Fourier Transform (STFT). They conduct extensive investigations while the elevator runs at different speeds, in different directions and with varying weights. The results show that applying principal component analysis is very helpful for quickly and effectively monitoring the condition of a traction machine in different scenarios. Overall, this method can help ensure that elevator traction machines are operating safely.

Yun et al. [8] proposed an effective vision-based method for recognizing objects that involves two main stages. In the first stage, a lightweight semantic segmentation neural network called ENet is used to extract the Region of Interest (ROI). This allows the objects that are not of interest or that are part of the background to be masked out. The areas that have the potential to contain objects of interest are the only regions that will be recognized. In the second stage, the masked image from the first stage is processed through the YOLO neural network to achieve efficient and accurate recognition. While the results from the first stage may not be perfect, the second stage can still achieve high accuracy. Experiments on embedded devices have demonstrated that using this two-stage method for object recognition not only saves power and computation but also significantly improves accuracy.

High-speed trains rely on a pantograph to provide power by connecting to the power-lines. The pantograph's status is critical to the functioning of the high-speed railway (HSR). To detect and locate the pantograph in images captured by a specific camera, Tan et al. [9] developed a dedicated detection method that uses YOLOv4. They trained this model on data collected from real-world scenarios. Since the camera that watches the pantograph is mounted outside of the train, it can be susceptible to various types of interference, such as rainwater-induced blurring or dirt on the camera lens. To better understand the health status of the camera and analyze the interference affecting the performance of YOLOv4 detection, a classification method is proposed. This method counts the number of blobs appearing in the image to determine if the camera is affected by dirt or blur. In addition, since the image backgrounds of the photograph can be diverse in different scenarios and can significantly affect the detection performance of YOLOv4, a method was developed to infer the categories of complex backgrounds. Overall, this proposed system provides an effective and efficient way to detect and locate the pantograph on high-speed trains, despite the challenges posed by environmental factors.

Ren et al. [10] proposed a complete system that enables forklifts to transfer pallets accurately and efficiently in warehouse environments. This system has three main components: pallet monitoring using an RGB surveillance camera, pallet positioning with an RGB-D camera mounted on the forklift, and a dedicated controller that instructs the forklift to manipulate the pallet with high precision. To detect pallets that are far away from the cameras in the pallet monitoring module, a transformer-based prediction head is incorporated into the YOLOv5 network. This allows for the detection of small targets that correspond to only a few pixels. For pallet positioning, deep feature maps are generated from the RGB-D images and fed into a 3D key points detection network. This network accurately detects the eight corner points of the pallets' two square apertures. By fitting the extracted key points, the pose of the pallet relative to the forklift can be determined. Once the pose of a pallet is known, the forklift is controlled to transfer the pallet with a trajectory controller that incorporates forklift motion cycle prediction into the control process. The proposed system has been extensively tested in real-world warehouse scenarios and has been shown to be effective and reliable.

3. Conclusions

This Special Issue encompasses a diverse array of topics related to efficient sensing and intelligence for embedded devices. The advancements showcased in the ten research papers published within this Special Issue highlight the significance of devising practical and theoretical technologies to tackle challenges posed by resource-constrained computation platforms and adverse external interference across various applications, including robotics, manufacturing, transportation, and energy systems. These cutting-edge solutions strive to enhance the efficiency, accuracy, and robustness of embedded sensing and intelligence while surmounting physical limitations and diverse environmental obstacles. The research papers in this collection contribute to the progress of efficient intelligence technologies, offering valuable insights and inspiration for both academic researchers and industry practitioners in their quest to develop more advanced and efficient embedded sensing systems.

Author Contributions: Conceptualization, X.Z. and Y.L.; formal analysis, X.Z.; investigation, X.Z. and Y.L.; resources, Y.L.; writing—original draft preparation, X.Z.; writing—review and editing, Y.L.; supervision, Y.L.; project administration, Y.L.; funding acquisition, Y.L. All authors have read and agreed to the published version of the manuscript.

Conflicts of Interest: The authors declare no conflict of interest.

References

1. Huang, T.; Chen, Z.; Gao, W.; Xue, Z.; Liu, Y. A USV-UAV Cooperative Trajectory Planning Algorithm with Hull Dynamic Constraints. *Sensors* **2023**, *23*, 1845. [CrossRef] [PubMed]
2. Chen, C.; Ma, Y.; Lv, J.; Zhao, X.; Li, L.; Liu, Y.; Gao, W. OL-SLAM: A Robust and Versatile System of Object Localization and SLAM. *Sensors* **2023**, *23*, 801. [CrossRef] [PubMed]
3. Tian, X.; Yi, P.; Zhang, F.; Lei, J.; Hong, Y. STV-SC: Segmentation and Temporal Verification Enhanced Scan Context for Place Recognition in Unstructured Environment. *Sensors* **2023**, *22*, 8604. [CrossRef] [PubMed]
4. Wu, Q.; Meng, Q.; Tian, Y.; Zhou, Z.; Luo, C.; Mao, W.; Zeng, P.; Zhang, B.; Luo, Y. A Method of Calibration for the Distortion of LiDAR Integrating IMU and Odometer. *Sensors* **2022**, *22*, 6716. [CrossRef] [PubMed]
5. Xia, Z.; Wang, Y.; Ma, L.; Zhu, Y.; Li, Y.; Tao, J.; Tian, G. A Hybrid Prognostic Method for Pro-ton-Exchange-Membrane Fuel Cell with Decomposition Forecasting Framework Based on AEKF and LSTM. *Sensors* **2022**, *23*, 166. [CrossRef] [PubMed]
6. Lin, G.; Liu, K.; Xia, X.; Yan, R. An Efficient and Intelligent Detection Method for Fabric Defects based on Improved YOLOv5. *Sensors* **2022**, *23*, 97. [CrossRef] [PubMed]
7. Li, D.; Yang, J.; Liu, Y. Research on State Recognition Technology of Elevator Traction Machine Based on Modulation Feature Extraction. *Sensors* **2022**, *22*, 9247. [CrossRef] [PubMed]
8. Yun, H.; Park, D. Efficient Object Detection Based on Masking Semantic Segmentation Region for Lightweight Embedded Processors. *Sensors* **2022**, *22*, 8890. [CrossRef] [PubMed]

9. Tan, P.; Cui, Z.; Lv, W.; Li, X.; Ding, J.; Huang, C.; Ma, J.; Fang, Y. Pantograph Detection Algorithm with Complex Background and External Disturbances. *Sensors* **2022**, *22*, 8425. [CrossRef] [PubMed]
10. Ren, J.; Pan, Y.; Yao, P.; Hu, Y.; Gao, W.; Xue, Z. Deep Learning-Based Intelligent Forklift Cargo Accurate Transfer System. *Sensors* **2022**, *22*, 8437. [CrossRef] [PubMed]

Article

A USV-UAV Cooperative Trajectory Planning Algorithm with Hull Dynamic Constraints

Tao Huang [1,2], Zhe Chen [1,2], Wang Gao [3], Zhenfeng Xue [1,2,*] and Yong Liu [1,2,*]

[1] Institute of Cyber-Systems and Control, Zhejiang University, Hangzhou 310027, China
[2] Intelligent Perception and Control Center, Huzhou Institute of Zhejiang University, Huzhou 313098, China
[3] Science and Technology on Complex System Control and Intelligent Agent Cooperation Laboratory, Beijing 100191, China
* Correspondence: zfxue0903@zju.edu.cn (Z.X.); yongliu@iipc.zju.edu.cn (Y.L.)

Abstract: Efficient trajectory generation in complex dynamic environments remains an open problem in the operation of an unmanned surface vehicle (USV). The perception of a USV is usually interfered by the swing of the hull and the ambient weather, making it challenging to plan optimal USV trajectories. In this paper, a cooperative trajectory planning algorithm for a coupled USV-UAV system is proposed to ensure that a USV can execute a safe and smooth path as it autonomously advances through multi-obstacle maps. Specifically, the unmanned aerial vehicle (UAV) plays the role of a flight sensor, providing real-time global map and obstacle information with a lightweight semantic segmentation network and 3D projection transformation. An initial obstacle avoidance trajectory is generated by a graph-based search method. Concerning the unique under-actuated kinematic characteristics of the USV, a numerical optimization method based on hull dynamic constraints is introduced to make the trajectory easier to be tracked for motion control. Finally, a motion control method based on NMPC with the lowest energy consumption constraint during execution is proposed. Experimental results verify the effectiveness of the whole system, and the generated trajectory is locally optimal for USV with considerable tracking accuracy.

Keywords: USV-UAV cooperation; trajectory generation; under-actuated constraint; numerical optimization; hull dynamics

Citation: Huang, T.; Chen, Z.; Gao, W.; Xue, Z.; Liu, Y. A USV-UAV Cooperative Trajectory Planning Algorithm with Hull Dynamic Constraints. *Sensors* **2023**, *23*, 1845. https://doi.org/10.3390/s23041845

Academic Editor: Sašo Blažič

Received: 10 November 2022
Revised: 19 January 2023
Accepted: 2 February 2023
Published: 7 February 2023

1. Introduction

Unmanned surface vehicles (USVs) are a kind of specific ships with the ability of autonomous mission execution, which are widely used in various applications, including marine resource exploration, water resource transportation, patrol and defense in key areas and river regulation [1,2]. Progress has been made in a large number of research areas, including environmental perception [3,4], formation control [5,6], navigation [7,8], and so on. Environmental perception and trajectory generation are the two most important techniques when the USVs are executing in unknown environments. In particular, when the environment contains dynamic obstacles, USVs struggle to achieve accurate trajectory planning and tracking due to the lack of effective obstacle information. As a result, the autonomous navigation system may fail.

During the navigation process of a USV, the sensing devices, such as radar or camera, are located at a low observation point, which is detrimental to environmental perception because the adjacent obstacles in the front and behind will block each other. Moreover, the input of the sensors often contains noise caused by hull shaking on the water. This makes precise environmental perception a difficult problem for USVs and affects the success rate of trajectory generation. Usually, simultaneous localization and mapping (SLAM) [9] technology is required to construct the global map. However, this kind of method requires a huge computational load, and it is intractable to deal with dynamic objects in the water environment.

A feasible solution is to design a USV-UAV cooperative system to tackle the above problems, where the unmanned aerial vehicle (UAV) plays the role as a flying sensor. As shown in Figure 1, the USV has long cruise capability, but its perception is disturbed and limited by the circumstance. Hence the UAV flies over the USV, providing more stable and comprehensive information. Semantic segmentation [10,11] and 3D projection are used in this paper to transfer obstacle information in the field of vision of the UAV to the coordinate system of the USV. Semantic segmentation extracts pixel information of environmental obstacles, and a camera projection model helps to transfer the pixel information to 3D information. By doing this, global map information around the USV can be obtained efficiently and in real-time, implying the USV-UAV cooperative system can improve the perception ability of the USV effectively, allowing the USV to perform tasks in more complex water circumstances.

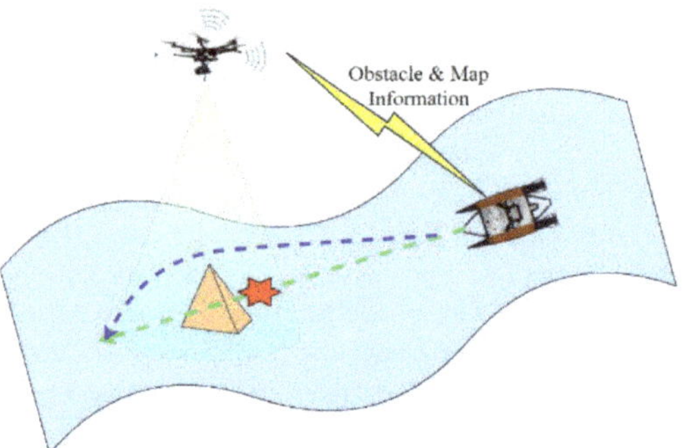

Figure 1. An illustration of the USV-UAV cooperative system, where the UAV provides wide obstacles and map information to guide the USV to generate an obstacle avoidance trajectory.

An initial obstacle avoidance trajectory is firstly generated by a graph-based search method [12]. However, such a method was originally designed for path searching on vast geographical scenarios, which does not consider the USV's dynamic characteristics. On the other hand, USV is famous for its under-actuated motion characteristics [13], which makes it hard to be controlled well, even when an optimal trajectory is planned. In this paper, we design a numerical optimization method to optimize the trajectory. Specifically, we take the hull dynamic constraints into account when modeling the optimization problem. As a result, the generated trajectory not only allows the obstacle avoidance rule, but also fits the motion characteristics of a USV. This makes the generated trajectory easier to be tracked under the same control conditions.

Finally, a control method with the lowest energy consumption per execution task is designed under a new numerical optimization problem. It ensures that the power consumption is optimal when the USV is actuated to track the given optimal trajectory, which is a very useful technique in real-world applications. The performance of the trajectory generation and tracking is comprehensively compared and analyzed in the simulated environments, and it verifies the effectiveness of our proposed novel framework.

In summary, the contributions of this paper are listed as follows.

- A novel USV-UAV cooperative system is proposed, where the UAV acts as a flying sensor to provide global map information around the USV by semantic segmentation and 3D projection, providing more comprehensive and effective perception results for navigation planning.
- A numerical optimization problem is formulated during the trajectory generation process. It considers the hull under-actuated dynamic constraints and perception of the UAV, which can generate a fuel-saving trajectory in real-time optimization.

- The lowest energy consumption control law is proposed to track the generated trajectory efficiently and accurately, and extensive experiments are conducted to verify the effectiveness of the USV-UAV cooperative system.

2. Related Works

2.1. Trajectory Planning for USV

Trajectory planning aims to automatically generate an obstacle avoidance trajectory for a USV when the local or global map is given. Among existing methods, the mainstream trajectory planning methods are mainly divided into two categories, i.e., path search and trajectory generation.

There are two research directions for the path search methods, including graph search and random sampling. Typical graph search methods include the A* [14] and Dijkstra [15] algorithm, as well as their derivatives [16]. These methods mainly discretize the known map into interconnected grids and find the shortest path according to the heuristic parameters. The disadvantage of this kind of method is that the search dimension in the large map is expanding, and the calculation time shows a rapid upward trend. Among random sampling methods, typical varieties include RRT [17] and its derivatives [18], which dynamically find feasible paths by randomly sampling points in the map and constructing random exploratory trees. The method can show better performance for large maps, but its shortcomings are also very obvious. It is easy to be guided to a locally optimal solution, and it is difficult to generate feasible paths in narrow areas when system's computing resources are limited. The common problem of the above methods is that the generated path curvature is discontinuous, and trajectory smoothing is needed afterward.

For the trajectory generation methods, curve interpolation methods, such as B-spline [19], are commonly used to smooth the trajectory. The smoothness of the trajectory and motion state is guaranteed by the continuity theorem of higher-order derivatives of a curve. Meanwhile, numerical optimization methods are also widely used, such as minimum snap [20] and near-optimal control [21].

Some methods can also combine path search with trajectory generation, such as domain reduction-based RRT* [22] and Hybrid A* [23]. In this paper, the proposed method belongs to the numerical optimization method . It adds the dynamic and kinematic constraints of unmanned craft in the trajectory generation part so that the generated trajectory is more in line with the dynamic characteristics of the hull.

2.2. The USV-UAV Cooperative System

With the rapid development of automation and artificial intelligence technology, unmanned aerial vehicle (UAV) technology has made significant progress in recent years. Compared with USV, the advantage of UAV is that it has a broader field of vision and faster movement speed and can provide more comprehensive and effective data information for USV. In addition, UAV has the advantages of flying height and that its communication ability is less affected by the environment. It can be used to provide communication relay services for multiple USVs located in different positions. Due to the strong complementarity between USV and UAV in perception, communication, operation time, and other aspects, researchers have focused on the coordination of having UAV serve USV and have successfully verified that this method can effectively solve the problem mentioned above of self-awareness of a USV. Ref. [24] focused on the search and rescue of USVs in flood scenes and proposed a collaborative mode of manipulating a UAV to establish the global map first, providing complete map information and target localization for subsequent USV planning. Ref. [25] proposed a cooperative formation control algorithm for a single USV and multiple UAVs. The method is based on the leader-follower distributed consensus model, and the position and orientation of each boat are determined by the RGB image color-space features acquired by the UAV camera. Ref. [26] considered the strong search capability of the UAV in the air, combined with the actual target strike capability of the USV, and proposed a two-stage cooperative path planning algorithm on the water and underwater based on the particle swarm optimization algorithm. Ref. [27] proposed an effective game incentive mechanism for the task assignment problem in the cooperative operation of USVs and UAVs, which reduced the task cost and improved the task efficiency. Ref. [28] proposed that the LVS-LVA framework to be applied the cooperative motion control of USV-UAV.

Although, most of these methods are cooperative ways to provide UAV environmental data and perceptual information for the navigation task of a USV. With the development of computer vision technology, the accuracy and robustness of the perception algorithm they use need to be improved. In addition, they did not consider the trajectory of the USV and its tracking control link, and the proposed collaborative framework can not be fully applied to the autonomous navigation task of USVs.

3. Cooperative Trajectory Generation

In the USV-UAV cooperative system, the USV has a stable environmental self-supporting ability, and the UAV is flexible and environmentally adaptable. In the process of autonomous navigation of the USV, relying on the wide field of vision and strong environmental perception provided by the UAV, it can generate a more reasonable trajectory and skillfully avoid various kinds of obstacles.

3.1. Environmental Perception and 3D Projection

Environmental perception is vital when the USV is performing in unknown water areas. Different observation angles have a significant influence on the observed results. As shown in Figure 2, the USV and UAV have different angles of view. The USV observes the environment from a horizontal perspective, which may lead to serious visual occlusion, whereas the UAV performs environmental perception from a top-down perspective, which enables more accurate map-view information.

(a) USV angle of view   (b) UAV angle of view

Figure 2. Perspective difference between USV and UAV.

Concerning the accuracy of obstacle recognition and the calculation efficiency, we use semantic segmentation technology [29,30] based on deep learning to extract pixel-level obstacle information from the image data obtained by the UAV's camera. For a given image, the position, shape and size of the obstacles in the environment can be judged by assigning each pixel with a two-categorical label: '0' indicates a safety area and '1' denotes an area in which the obstacles are located.

In this paper, we use DeepLab [10] as the semantic segmentation network and replace the backbone with MobileNet [31]. On the one hand, it reduces the amount of computation. On the other hand, in the process of feature extraction, with the help of the atrous spatial pyramid pooling (ASPP) module, it can effectively improve the global receptive field and the recognition effect. The overall network architecture is illustrated in Figure 3.

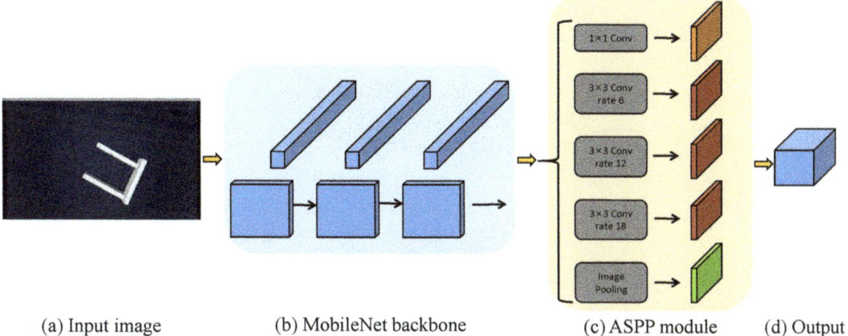

(a) Input image   (b) MobileNet backbone   (c) ASPP module   (d) Output

Figure 3. The network architecture of the semantic segmentation algorithm deployed on the UAV.

After obtaining the pixel coordinates of obstacles in the image, it needs to convert the obstacle coordinate information into a unified global coordinate. We define the coordinate system of the UAV as U, the camera coordinate system as C, and the global coordinate system as G. Thus the transformation from U to C can be represented by $T_{UC} = [R|T] \in R^{4 \times 4}$, where R is the rotation matrix and T is the translation matrix. $T_{GU} \cdot T_{UC}$ denotes the transformation matrix from G to C. Assuming that the coordinates of the obstacle point m in the pixel coordinate system are (u, v), according to the imaging principle of the pinhole camera model, the relationship between its position in the camera coordinate system can be expressed as

$$
\begin{cases}
u = f_x \cdot \dfrac{x}{z} + c_x \\
v = f_y \cdot \dfrac{y}{z} + c_y,
\end{cases}
\tag{1}
$$

where f_x and f_y denote the focal length in the x and y direction and c_x and c_y are the positions of the origin of the image plane, which can usually be regarded as the center of the image. Thus, the relationship between the 3D points in the global coordinate system $M = (x, y, z)$ and the pixel coordinate system $m = (u, v)$ is denoted by

$$
s \cdot
\begin{bmatrix} u \\ v \\ 1 \\ 1 \end{bmatrix}
=
\begin{bmatrix} f_x & 0 & c_x \\ 0 & f_y & c_y \\ 0 & 0 & 1 \\ 0 & 0 & s \end{bmatrix}
\cdot T_{GU} \cdot T_{UC} \cdot
\begin{bmatrix} x \\ y \\ z \\ 1 \end{bmatrix},
\tag{2}
$$

where s is the scaling factor, which can be regarded as the depth information of each pixel. In this paper, a binocular camera carried by the UAV is used to obtain the pixel depth s. Through this way of 3D coordinate projection, the pixel information sensed by the UAV in real-time can be projected into the global coordinate system, forming the 3D perception ability of the USV to the environment.

3.2. Initial Trajectory Generation

In order to generate an obstacle avoidance trajectory, this paper applies the Hybrid A* algorithm [23] to provide an initial path, as shown in Algorithm 1. Given the initial state of the USV ($s = (x_0, y_0, \varphi_0)$) and the navigation target state ($e = (x_f, y_f, \varphi_f)$), the algorithm first puts the initial state into the open list. Then it iteratively reads the node with the lowest cost in the open list as the current parent node, and generates the next child node according to the current node state, system motion mode and obstacle map. Unlike the A* algorithm, the Hybrid A* algorithm adds the orientation dimension to the coordinate system. Therefore, the criteria for reaching the target state is that the distance between the coordinates of the node and the target point is less than the threshold of the reaching distance, and the collision-free Reeds–Shepp curve can be generated through the node state and the target point state.

Algorithm 1 Trajectory Search with Hybrid A*

Input: x_0, x_f, map
Output: Trajectory T
1: Function Search(x_0, x_f, map)
2: $open \leftarrow \phi$, $close \leftarrow \phi$
3: $open$.push(x_0)
4: **while** $open$ is not ϕ **do**
5:     $x_n \leftarrow open$.pop()
6:     $close$.push(x_n)
7:     **if** x_n.near(x_f) **then**
8:         **if** reedsheep(x_n, x_f) **then**
9:             **return** $path(x_f)$
10:    **else**
11:         **for** $x_{succ} \in$ successor(x_n) **do**
12:             **if** x_{succ}.safe() and not exist(x_n, $close$) **then**
13:                 $g \leftarrow g(x_n) + g(x_{succ}, x_n)$
14:                 **if** not exist(x_{succ}, $open$) or $g < g(x_{succ})$ **then**
15:                     pred(x_{succ}) $\leftarrow x_n$
16:                     h(x_{succ}) $\leftarrow$ Heuristic(x_{succ}, x_f)
17:                     **if** not exist(x_{succ}, $open$) **then**
18:                         $open$.push(x_{succ})
19:                     **else**
20:                         $open$.rewrite(x_{succ})
21: **return** $null$

---

4. Trajectory Optimization and Tracking

The USV is an under-actuated robot operation system where the number of control variables of the system is less than the degrees of freedom of the system. In the trajectory optimization process, if the dynamic constraints of this under-actuated characteristic are added to the optimization process, an optimal trajectory more in line with the characteristics of ship motion can be generated.

4.1. Trajectory Optimization with Dynamics

The motion model of the USV is a mathematical model with 6 degrees of freedom when it is complete. For simplicity, we can ignore the motion of the hull in the heave, roll and pitch directions, and simplify it into a 3-degrees of freedom with surge, sway and yaw, represented by x, y and φ. The mathematical expression of the hull dynamics can be expressed as

$$\begin{cases} \dot{\eta} = J(\eta)v \\ M\dot{v} = \tau - C(v)v - Dv, \end{cases} \tag{3}$$

where $\eta = (x, y, \varphi) \in R^{3 \times 1}$ denotes the state variables, and $v = (u, v, r) \in R^{3 \times 1}$ denotes the speed variables. $J \in R^{3 \times 3}$ is the transition matrix, and $C \in R^{3 \times 3}$ is the Coriolis centripetal force matrix. $M \in R^{3 \times 3}$ is the inertial matrix, and $D \in R^{3 \times 3}$ is the damping matrix. $\tau = (\tau_u, 0, \tau_r) \in R^{3 \times 1}$ is the thrust matrix. For a catamaran, the thrust matrix can be expressed as

$$\begin{cases} \tau_u = T_1 + T_2 \\ \tau_r = (T_1 - T_2) \cdot B, \end{cases} \tag{4}$$

where T_1 and T_2 are the thrusts of two propellers, and B is their distance. The USV can be viewed as a linear time-invariant (LTI) system. Its state variables X and control variable τ can be represented by

$$\begin{cases} X = [x, y, \varphi, u, v, r]^T \\ \tau = [\tau_u, 0, \tau_r]^T. \end{cases} \tag{5}$$

The system dynamics are as follows

$$\begin{cases} \dot{x} = u\cos(\varphi) - v\sin(\varphi) \\ \dot{y} = u\sin(\varphi) + v\cos(\varphi) \\ \dot{\varphi} = r \\ m_{11}\dot{u} - m_{22}ur + d_{11}u = \tau_u \\ m_{22}\dot{v} - m_{11}ur + d_{22}v = 0 \\ m_{33}\dot{r} + (m_{22} - m_{11})uv + d_{33}r = \tau_r. \end{cases} \tag{6}$$

Based on Hybrid A*, the global trajectory is optimized twice with the following constraints, including position, velocity, angular velocity and control input, as well as waypoint state constraints. The reference waypoint state is the sub-optimal trajectory obtained by considering the vehicle model, which can only provide the simulated optimal information of obstacle avoidance, heading speed and other controls. In this paper, we consider the state vector error in the optimization objective function as a soft constraint. The final optimization objective can be represented as

$$min \quad \frac{1}{2}\{\sum_{i=0}^{N}[(X_i - X_i^{ref})^T W_x (X_i - X_i^{ref}) + \tau_i^T W_\tau \tau_i] + \sum_{i=1}^{N}(\tau_i - \tau_{i-1})^T W_u (\tau_i - \tau_{i-1})\}, \tag{7}$$

where X_i^{ref} denotes the reference state variables generated by Hybrid A*, and $W_x = diag\{50, 50, 20, 15, 15, 15\}$, $W_\tau = diag\{5, 0, 5\}$ and $W_u = diag\{3, 0, 3\}$ represent the positive definite, cost and weight matrices, respectively. Moreover, to ensure adequate accuracy in the trajectory, we choose 0.05 s as the sampling period.

We adopt the methods of minimizing the control quantity and minimizing the continuous control difference to ensure that the global trajectory generated by optimization can take into account the trajectory index factors, such as the smoothing of the control quantity and the minimization of the energy consumption at the same time. The overall algorithm flow is shown in Algorithm 2.

Algorithm 2 Global Trajectory Optimization

Input: X_0, X_f, *path*
Output: X
  1: Function OptiTraj(X_0, X_f, *path*)
  2: **for** $i = 0$ to N **do**
  3:      **if** $i == 1$ **then**
  4:          $X(i) = X_0$
  5:      **else if** $i == $ N **then**
  6:          $X(i) = X_f$
  7:      **else**
  8:          $X(i).x = path_i.x$
  9:          $X(i).y = path_i.y$
10:          $X(i).\varphi = path_i.\varphi$
11: Set constraints C
12: Set Objective Function J
13: Optimize(J, *path*, C, X)
14: **return** X

4.2. Tracking Control with NMPC

Nonlinear model predictive control (NMPC) [32] is famous for its ability to improve local tracking precision. It performs periodic real-time optimization according to the prediction time window to achieve the purpose of iterative control to reduce tracking error. Through the numerical optimization algorithm proposed above, the global trajectory based on the kinematic and dynamic constraints of the USV can be obtained, in which the reference control quantity can be obtained. Therefore, the trajectory optimization uses the error index of control quantity as the optimization target. Setting the current time as t_j and the prediction time window as W_n, the optimization problem in terms of NMPC can be formulated as

$$min \quad \frac{1}{2} \sum_{i=t_j}^{t_j+W_n} [(X_i - X_i^{ref})^T W_{mpcx}(X_i - X_i^{ref}) + (\tau_i - \tau_i^{ref})^T W_{mpc\tau}(\tau_i - \tau_i^{ref})$$
$$+ (\tau_i - \tau_{i-1})^T W_{mpcu}(\tau_i - \tau_{i-1})], \tag{8}$$

where the first term represents the error between the state variable and the reference state variable, which is mainly used to improve the accuracy of state tracking and maintenance in the process of real-time control. The second term represents the error between the control variable and the reference control variable. This term is used to meet the index of the lowest energy consumption. Although this problem has been considered in detail in the context of optimization objectives in global trajectory planning, secondary planning in local tracking control can achieve better results. The third term can improve the smoothness of input variables in actual control and meet the needs of practical application control. $W_{mpcx} = diag\{10, 10, 4, 2, 2, 2\}$, $W_{mpc\tau} = diag\{2, 0, 2\}$ and $W_{mpcu} = diag\{4, 0, 4\}$ represent the positive definite, cost and weight matrices, respectively. And considering the control requirements of real-time operation and stability, we choose W_n to be 30, 0.05 s as the sampling period, and the cycle of the NMPC algorithm call to be 0.1 s.

5. Experimental Analysis

In this section, we perform simulation experiments using the open source Otter USV simulator [33] within the ROS environment. The Otter USV simulator is a catamaran 2.0-m long, 1.08-m wide and 1.06-m high. When fully assembled, it weighs 65 kg, and has the ability to be disassembled into parts weighing less than 20 kg, such that a single operator can launch the Otter from a jetty, lake, beach or riverbank. A PX4 drone autopilot is used as the UAV, which is mounted with a monocular camera. The Otter USV is traveling within a 200×100 square meter area, with many blocks placed therein as obstacles. We set up several different obstacle terrains to test the crossing ability of the USV-UAV cooperative system.

5.1. Obstacle Recognition Ability

Firstly, we perform experiments on the ability of obstacle recognition by the USV monocular camera. The semantic segmentation algorithm is used to recognize objects. Several terrains are randomly placed in the virtual environment. Some of the segmentation results are shown in Figure 4, from which we can see that the proposed light-weight segmentation network can successfully identify obstacles in the environment. Although there are some empty areas in the middle or on the edges of the obstacles, the basic shape of the obstacles has been preserved. In the post-processing stage, image expansion can be used to increase the safe collision avoidance area and ensure the reliability of navigation. After that, 3D projection can be performed to convert the pixel information into 3D information in a global coordinate system.

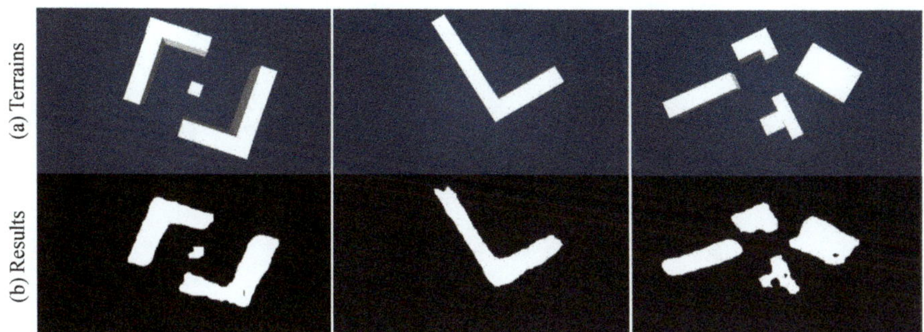

Figure 4. Obstacle recognition results of different terrains.

5.2. Trajectory Generation Performance

The trajectory generation result is illustrated in Figure 5, from which we can see that the generated trajectory not only meets the collision avoidance condition, but also conforms to the hull's kinematic characteristics. In the experiment, the Otter is an under-actuated USV and cannot provide direct lateral thrust during its operation. This requires that the running trajectory of the USV must be smooth enough, because too many bends will bring instability to the motion control of the USV and lead to the failure of path trajectory. The corresponding results can be seen in the subsequent path tracking control experiments.

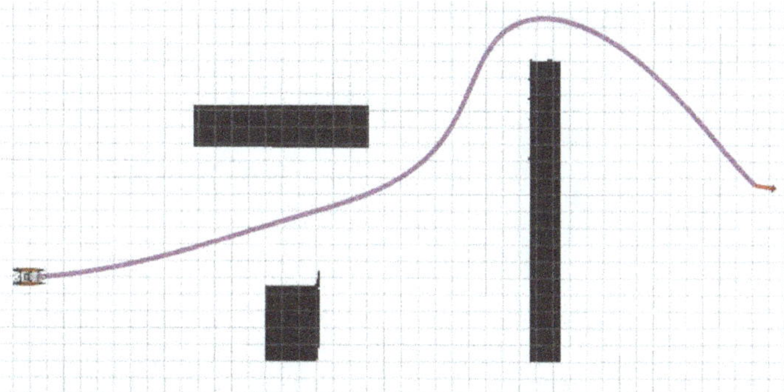

Figure 5. Global trajectory generation performance of the USV-UAV cooperative system.

The changing trend of the state and control quantity of the USV with time for the generated trajectory can be found in Figure 6. Overall, the quantities show a relatively gentle trend, especially for the x and y quantities, which verifies the smoothness of the trajectory. Higher order quantities such as u, v and yaw also present a gentle trend. Those are sufficient to show the effectiveness of the trajectory optimization method.

We also performed an ablation study on the proposed method. As shown in Figure 7, the LOP and GP+LOP methods are compared. LOP denotes the trajectory generation with local optimization planning, which means the global map provided by the UAV is unknown. Due to the limited perception field of the USV, it will take action to perform local trajectory planning unless it is near the obstacle. GP+LOP denotes global planning without trajectory optimization, which means the global map is known while trajectory optimization is not performed. Without the optimization stage, the generated trajectory shows a twisted shape, which is not optimal. GOP+LOP denotes the proposed method. In the lower left-corner of each sub-figure, the total length of the generated trajectory is shown. Our method obtains the shortest planned path with the best smoothness.

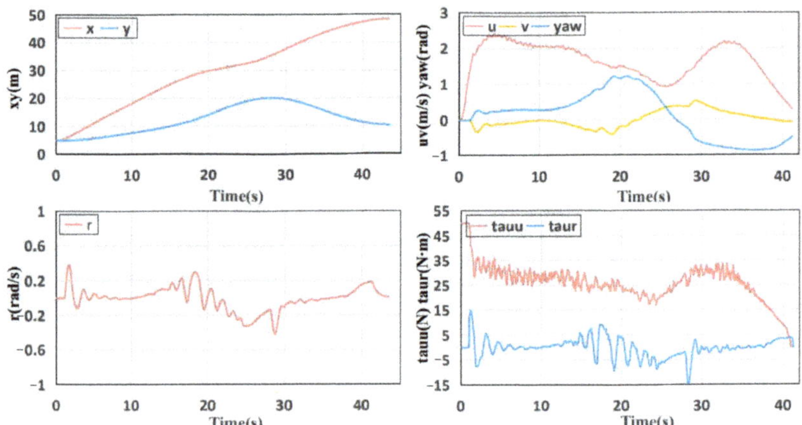

Figure 6. The changing trend of the state and control quantity of the USV with time.

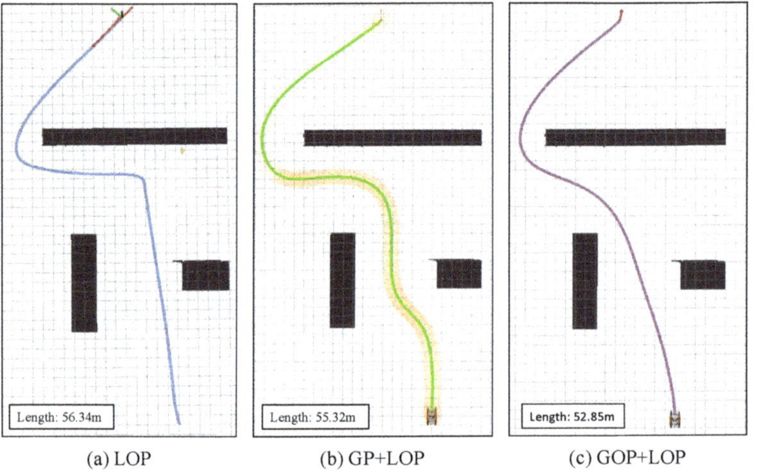

(a) LOP　　　　　　　　(b) GP+LOP　　　　　　　　(c) GOP+LOP

Figure 7. Trajectory generation comparison with different methods. LOP: trajectory generation with local optimization planning (global map provided by the UAV is unknown); GP+LOP: global planning without trajectory optimization; and GOP+LOP: the proposed method.

Here, we also compare the three methods quantitatively in Table 1. The indexes, such as RMSE, max error, speed and time, are evaluated by driving the hull to move. With the trajectory optimization method, the generated trajectory is more in line with the kinematic characteristics of the hull. As such, the tracking error, execution speed and control time achieve optimal values compared with other methods.

Table 1. Quantitative comparison of different trajectory generation methods.

Method	Length (m)	RMSE (m)	Max Error (m)	Speed (m/s)	Time (s)
LOP	56.34	0.120	0.3045	1.513	0.0667
GP+LOP	55.32	0.118	0.3047	1.608	0.0697
GOP+LOP	52.85	0.113	0.2312	1.675	0.0506

5.3. Tracking Control Performance

To further verify the effectiveness of the proposed NMPC tracking control module, extensive comparative experiments are conducted. As shown in Figure 8, GOP+LP denotes the tracking control method without optimization, i.e., the plain PID with adjusted parameters. The proposed NMPC shows better tracking control performance qualitatively and quantitatively. There is no prediction time window for GOP+LP, so there will be many minor adjustments, resulting in an actual motion trajectory that is not smooth.

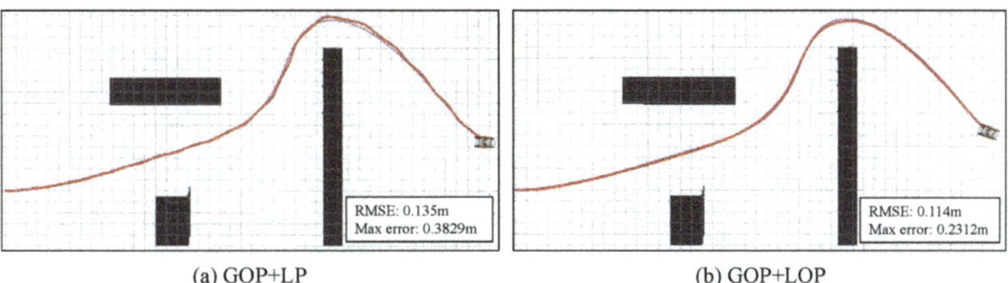

Figure 8. Tracking control performance comparison. GOP+LP denotes the tracking control method without optimization, i.e., the PID control. GOP+LOP denotes the proposed method with NMPC control.

The execution states of different tracking control methods are visualized in Figure 9, from which the plain PID control shows unstable tracking states. Especially for the control input, the τ_r shows a divergent trend, which may lead to the input variable exceeding the controllable range and adversely affecting the motion control of the USV.

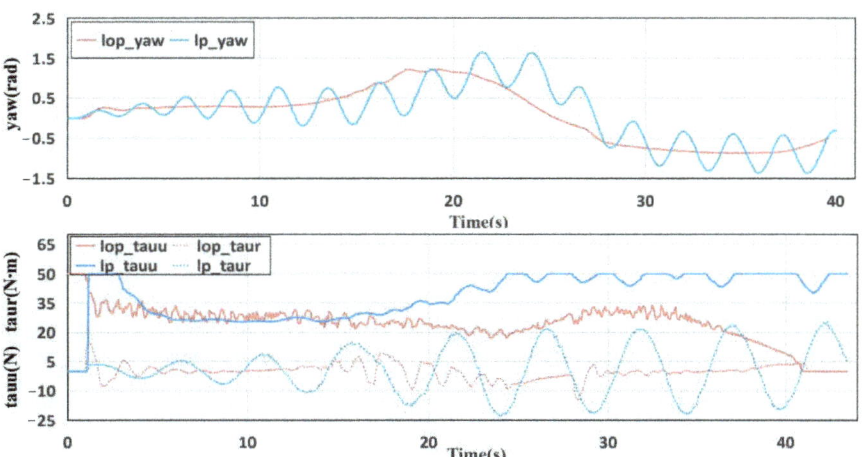

Figure 9. Execution state comparison of motion tracking control.

The quantitative comparison of tracking control methods can be found in Table 2, from which the proposed method shows better performance than GOP+LP (i.e., plain PID control). The proposed method not only achieves a smaller tracking control error, but also drives the USV at a quicker speed. Those particularly prove the effectiveness of the combination of motion control and trajectory generation with hull dynamics.

Table 2. Quantitative comparison of tracking control methods.

Method	RMSE (m)	Max Error (m)	Speed (m/s)
GOP+LP	0.135	0.3829	1.327
GOP+LOP	0.113	0.2312	1.675

6. Conclusions

In this paper, a USV-UAV cooperative trajectory planning algorithm is proposed to overcome the problem of USV navigation in complex and multi-obstacle environments with an unknown global map. The proposed cooperative system is simple yet practical. In our method, the UAV acts as a flying sensor, providing a global map to the USV in real-time with semantic segmentation and 3D projection. Afterward, a graph search-based method is applied to generate an initial obstacle avoidance trajectory. An optimization method that considers the kinematic characteristics of the hull is proposed to make the trajectory more in line with the situation. Finally, an NMPC control method is applied to ensure high precision motion control of the USV. The proposed method has excellent performance and strong practicability in ocean engineering. In future work, we will verify the feasibility of the method in physical experiments and try to study the heterogeneous cooperation scheme of multi USV-UAV systems.

Author Contributions: Conceptualization, T.H., Z.C. and Z.X.; methodology, T.H.; software, Z.C.; validation, T.H. and Z.C.; formal analysis, Z.C.; writing—original draft preparation, T.H.; writing—review and editing, W.G. and Z.X.; supervision, Z.X. and Y.L.; project administration, Z.X. All authors have read and agreed to the published version of the manuscript.

Funding: This research received no external funding.

Institutional Review Board Statement: Not applicable.

Informed Consent Statement: Not applicable.

Data Availability Statement: Data will be made available upon request from the authors.

Conflicts of Interest: The authors declare no conflict of interest.

References

1. Wang, W.; Shan, T.; Leoni, P.; Fernández-Gutiérrez, D.; Meyers, D.; Ratti, C.; Rus, D. Roboat ii: A novel autonomous surface vessel for urban environments. In Proceedings of the 2020 IEEE/RSJ International Conference on Intelligent Robots and Systems (IROS), Las Vegas, NV, USA, 25–29 October 2020; pp. 1740–1747.
2. Chen, Z.; Huang, T.; Xue, Z.; Zhu, Z.; Xu, J.; Liu, Y. A Novel Unmanned Surface Vehicle with 2D-3D Fused Perception and Obstacle Avoidance Module. In Proceedings of the 2021 IEEE International Conference on Robotics and Biomimetics (ROBIO), Sanya, China, 27–31 December 2021; pp. 1804–1809.
3. Han, J.; Cho, Y.; Kim, J. Coastal SLAM with marine radar for USV operation in GPS-restricted situations. *IEEE J. Ocean. Eng.* **2019**, *44*, 300–309. [CrossRef]
4. Cheng, L.; Deng, B.; Yang, Y.; Lyu, J.; Zhao, J.; Zhou, K.; Yang, C.; Wang, L.; Yang, S.; He, Y. Water Target Recognition Method and Application for Unmanned Surface Vessels. *IEEE Access* **2021**, *10*, 421–434. [CrossRef]
5. Yan, X.; Jiang, D.; Miao, R.; Li, Y. Formation control and obstacle avoidance algorithm of a multi-USV system based on virtual structure and artificial potential field. *J. Mar. Sci. Eng.* **2021**, *9*, 161. [CrossRef]
6. Liu, H.; Weng, P.; Tian, X.; Mai, Q. Distributed adaptive fixed-time formation control for UAV-USV heterogeneous multi-agent systems. *Ocean Eng.* **2023**, *267*, 113240. [CrossRef]
7. Page, B.R.; DaRosa, J.; Lindler, J. USV Fleet Planning Considering Logistical Constraints Using Genetic Algorithm. In Proceedings of the OCEANS 2022, Hampton Roads, VA, USA, 17–20 October 2022; pp. 1–7.
8. Zou, X.; Xiao, C.; Zhan, W.; Zhou, C.; Xiu, S.; Yuan, H. A novel water-shore-line detection method for USV autonomous navigation. *Sensors* **2020**, *20*, 1682. [CrossRef] [PubMed]
9. Naus, K.; Marchel, Ł. Use of a weighted ICP algorithm to precisely determine USV movement parameters. *Appl. Sci.* **2019**, *9*, 3530. [CrossRef]
10. Chen, L.C.; Papandreou, G.; Kokkinos, I.; Murphy, K.; Yuille, A.L. Deeplab: Semantic image segmentation with deep convolutional nets, atrous convolution, and fully connected crfs. *IEEE Trans. Pattern Anal. Mach. Intell.* **2017**, *40*, 834–848. [CrossRef] [PubMed]

11. Yao, L.; Kanoulas, D.; Ji, Z.; Liu, Y. ShorelineNet: An efficient deep learning approach for shoreline semantic segmentation for unmanned surface vehicles. In Proceedings of the 2021 IEEE/RSJ International Conference on Intelligent Robots and Systems (IROS), Prague, Czech Republic, 27 September–1 October 2021; pp. 5403–5409.
12. Niu, H.; Savvaris, A.; Tsourdos, A.; Ji, Z. Voronoi-visibility roadmap-based path planning algorithm for unmanned surface vehicles. *J. Navig.* **2019**, *72*, 850–874. [CrossRef]
13. Fossen, T.I. *Handbook of Marine Craft Hydrodynamics and Motion Control*; John Wiley & Sons: Hoboken, NJ, USA, 2021.
14. Rana, K.; Zaveri, M. A-star algorithm for energy efficient routing in wireless sensor network. In *Trends in Network and Communications*; Springer: Berlin/Heidelberg, Germany, 2011; pp. 232–241.
15. Wang, H.; Yu, Y.; Yuan, Q. Application of Dijkstra algorithm in robot path-planning. In Proceedings of the 2011 Second International Conference on Mechanic Automation and Control Engineering, Hohhot, China, 15–17 July 2011; pp. 1067–1069.
16. Zheng, T.; Xu, Y.; Zheng, D. AGV path planning based on improved A-star algorithm. In Proceedings of the 2019 IEEE 3rd Advanced Information Management, Communicates, Electronic and Automation Control Conference (IMCEC), Chongqing, China, 11–13 October 2019; pp. 1534–1538.
17. Kuffner, J.J.; LaValle, S.M. RRT-connect: An efficient approach to single-query path planning. In Proceedings of the Proceedings 2000 ICRA. Millennium Conference. IEEE International Conference on Robotics and Automation. Symposia Proceedings (Cat. No. 00CH37065), San Francisco, CA, USA, 24–28 April 2000; Volume 2, pp. 995–1001.
18. Guo, W.; Tang, G.; Zhao, F.; Wang, Q. Global Dynamic Path Planning Algorithm for USV Based on Improved Bidirectional RRT. In Proceedings of the 32nd International Ocean and Polar Engineering Conference, Shanghai, China, 6–10 June 2022.
19. Zhang, X.; Wang, C.; Chui, K.T.; Liu, R.W. A Real-Time Collision Avoidance Framework of MASS Based on B-Spline and Optimal Decoupling Control. *Sensors* **2021**, *21*, 4911. [CrossRef] [PubMed]
20. Mellinger, D.; Kumar, V. Minimum snap trajectory generation and control for quadrotors. In Proceedings of the 2011 IEEE International Conference on Robotics and Automation, Shanghai, China, 9–13 May 2011; pp. 2520–2525.
21. Zhang, Y.; Li, S.; Weng, J. Learning and near-optimal control of underactuated surface vessels with periodic disturbances. *IEEE Trans. Cybern.* **2021**, *52*, 7453–7463. [CrossRef] [PubMed]
22. Wen, N.; Zhang, R.; Wu, J.; Liu, G. Online planning for relative optimal and safe paths for USVs using a dual sampling domain reduction-based RRT* method. *Int. J. Mach. Learn. Cybern.* **2020**, *11*, 2665–2687. [CrossRef]
23. Sedighi, S.; Nguyen, D.V.; Kuhnert, K.D. Guided hybrid A-star path planning algorithm for valet parking applications. In Proceedings of the 2019 5th International Conference on Control, Automation and Robotics (ICCAR), Beijing, China, 19–22 April 2019; pp. 570–575.
24. Ozkan, M.F.; Carrillo, L.R.G.; King, S.A. Rescue boat path planning in flooded urban environments. In Proceedings of the 2019 IEEE International Symposium on Measurement and Control in Robotics (ISMCR), Houston, TX, USA, 19–21 September 2019.
25. Xue, K.; Wu, T. Distributed consensus of USVs under heterogeneous uav-usv multi-agent systems cooperative control scheme. *J. Mar. Sci. Eng.* **2021**, *9*, 1314. [CrossRef]
26. Wu, Y. Coordinated path planning for an unmanned aerial-aquatic vehicle (UAAV) and an autonomous underwater vehicle (AUV) in an underwater target strike mission. *Ocean Eng.* **2019**, *182*, 162–173. [CrossRef]
27. Liu, J.; Su, Z.; Xu, Q. UAV-USV Cooperative Task Allocation for Smart Ocean Networks. In Proceedings of the 2021 IEEE 23rd Int Conf on High Performance Computing & Communications; 7th Int Conf on Data Science & Systems; 19th Int Conf on Smart City; 7th Int Conf on Dependability in Sensor, Cloud & Big Data Systems & Application (HPCC/DSS/SmartCity/DependSys), Haikou, China, 20–22 December 2021; pp. 1815–1820.
28. Li, J.; Zhang, G.; Li, B. Robust adaptive neural cooperative control for the USV-UAV based on the LVS-LVA guidance principle. *J. Mar. Sci. Eng.* **2022**, *10*, 51. [CrossRef]
29. Long, J.; Shelhamer, E.; Darrell, T. Fully Convolutional Networks for Semantic Segmentation. *IEEE Trans. Pattern Anal. Mach. Intell.* **2015**, *39*, 640–651.
30. Xue, Z.; Mao, W.; Jiang, W. Ehanet: Efficient hybrid attention network towards real-time semantic segmentation. In Proceedings of the 2020 IEEE 6th International Conference on Computer and Communications (ICCC), Chengdu, China, 11–14 December 2020; pp. 787–791.
31. Qin, Z.; Zhang, Z.; Chen, X.; Wang, C.; Peng, Y. Fd-mobilenet: Improved mobilenet with a fast downsampling strategy. In Proceedings of the 2018 25th IEEE International Conference on Image Processing (ICIP), Athens, Greece, 7–10 October 2018; pp. 1363–1367.
32. Magni, L.; Raimondo, D.M.; Allgöwer, F. *Nonlinear Model Predictive Control*; Lecture Notes in Control and Information Sciences; Springer: Berlin/Heidelberg, Germany, 2009; Volume 384.
33. Lenes, J.H. Autonomous Online Path Planning and Path-Following Control for Complete Coverage Maneuvering of a USV. Master's Thesis, NTNU, Trondheim, Norway, 2019.

Article

OL-SLAM: A Robust and Versatile System of Object Localization and SLAM

Chao Chen [1,†], Yukai Ma [1,†], Jiajun Lv [1], Xiangrui Zhao [1], Laijian Li [1], Yong Liu[1,*] and Wang Gao [2,*]

1   Institute of Cyber-Systems and Control, Zhejiang University, Hangzhou 310027, China; chenchao1924@zju.edu.cn (C.C.); yukaima@zju.edu.cn (Y.M.); lvjiajun314@zju.edu.cn (J.L.); xiangruizhao@zju.edu.cn (X.Z.); lilaijian@zju.edu.cn (L.L.)
2   Science and Technology on Complex System Control and Intelligent Agent Cooperation Laboratory, Beijing 100191, China
*   Correspondence: yongliu@iipc.zju.edu.cn (Y.L.); gaowang@iipc.zju.edu.cn (W.G.)
†   These authors contributed equally to this work.

Abstract: This paper proposes a real-time, versatile Simultaneous Localization and Mapping (SLAM) and object localization system, which fuses measurements from LiDAR, camera, Inertial Measurement Unit (IMU), and Global Positioning System (GPS). Our system can locate itself in an unknown environment and build a scene map based on which we can also track and obtain the global location of objects of interest. Precisely, our SLAM subsystem consists of the following four parts: LiDAR-inertial odometry, Visual-inertial odometry, GPS-inertial odometry, and global pose graph optimization. The target-tracking and positioning subsystem is developed based on YOLOv4. Benefiting from the use of GPS sensor in the SLAM system, we can obtain the global positioning information of the target; therefore, it can be highly useful in military operations, rescue and disaster relief, and other scenarios.

Keywords: SLAM; multi-sensor fusion; object tracking and localization

1. Introduction

For autonomous unmanned systems, SLAM technology can observe the surrounding stationary environment and build a 3D map of the environment through sensors such as cameras and LiDARs installed on the robot [1]. For dynamic scenes, SLAM-based 4D reconstruction technology can reconstruct 4D (3D+time) dynamic scenes with rigid moving objects [2–4]. However, the complexity of the actual scene means the SLAM system, with only positioning and mapping functions, is unable to meet the needs of many scenarios such as military operations, emergency rescue, and disaster relief. Besides, autonomous unmanned systems are often required to obtain positioning and environmental maps at the same time. It is necessary to develop an intelligent multifunctional perception system with self-positioning, mapping, target tracking, and positioning functions to search for objects of interest within the field of view and obtain the target location.

In order to improve the state estimation accuracy of SLAM systems, a large number of multi-sensor fusion methods have been used, such as the fusion of camera and IMU [5–8], LiDAR and IMU [9–11], and a combination of all of them [12–14]. The sensors used in these methods can be divided into local pose estimation sensors such as camera and IMU and global pose estimation sensors such as GPS and magnetometer. However, they all have their advantages and disadvantages, so the single use of a certain type of sensor limits the SLAM system in practical application [15–17]. The short-term results are more credible for local pose estimation sensors, but they have two shortcomings. One is that their pose estimation results do not have a global coordinate system, so the method is not reusable. The second is that when the system runs for a long time, there will be an inevitable cumulative drift. Although the loop closure detection method can correct the accumulated error in the SLAM system, problems such as difficult matching, a large amount of data,

Citation: Chen, C.; Ma, Y.; Lv, J.; Zhao, X.; Li, L.; Liu, Y.; Gao, W. OL-SLAM: A Robust and Versatile System of Object Localization and SLAM. *Sensors* **2023**, *23*, 801. https://doi.org/10.3390/s23020801

Academic Editor: Anastasios Doulamis

Received: 21 October 2022
Revised: 26 December 2022
Accepted: 26 December 2022
Published: 10 January 2023

and limited application scenarios still exist. The frequency of global pose estimation sensors is not high, so they cannot provide much continuous observation information. Furthermore, their measurement noise is relatively large; therefore, they cannot be directly used for pose estimation. However, they have global observation coordinates and are not affected by time accumulation. Therefore, fusing different sensors is an important method to enhance pose estimation accuracy.

For targeting, YOLOv4 [18] provides a high-speed and accurate target detection network model. The optical flow method and Kalman filtering are often used for target tracking. In terms of application, J.A. [19] proposes an automatic expert system, based on image segmentation procedures, that assists in safe landing through recognition and the relative orientation of the UAV and platform. Dr. Krishna [20] specified detection and tracking algorithms in terms of extracting the features of images and videos for security and scrutiny applications. We tend to fuse other sensors such as LiDAR or Radar to locate the target because monocular cameras lack depth information. Yifang [21] proposes the use of RADAR and Infrared sensor (IR) information for tracking and estimating target state dynamics. To project image-based object detection results and LiDAR-SLAM results onto a 3D probability map, Gong et al. [22] combine visual and range information into a frustum-based probabilistic framework.

For the above reasons, this paper proposes an online positioning, mapping, and target-tracking and location system based on camera, IMU, solid-state LiDAR, and GPS. Specifically, our SLAM system consists of the following four parts: a LiDAR-inertial subsystem (LIS), Visual-inertial subsystem (VIS), GPS-inertial subsystem (GIS), and global pose graph optimization (PGO). The LIS and VIS are tightly coupled, and there exists loose coupling between them. The combination of them can improve the accuracy and robustness of the whole system. Finally, the LIS and VIS results are sent to the PGO system for global pose graph optimization to eliminate accumulated drifts. Due to our distributed structure design of tightly coupled internal and loose coupled external subsystems, our system dramatically improved its robustness even in cases where one of the subsystems fails. In addition, a LiDAR–Camera fusion localization method is proposed based on conventional target detection and tracking. The global position of the target is obtained in real-time based on our SLAM.

To test the effectiveness of our system, we built the necessary sensor equipment and collected many scene-rich datasets, including high-altitude UAV aerial photography datasets, ground vehicle datasets, and ground handheld datasets. Considering that there are relatively few datasets that include the sensors we use, we open source all collected datasets for other researchers to use. Finally, we conduct extensive experiments on our dataset to test our system. Experiments show that our system can perform the expected function well with good accuracy and robustness.

The main contributions of this paper are as follows:

1. We propose a high-precision, high-robust multi-sensor fusion online SLAM system;
2. We propose an online target-tracking and localization system based on SLAM results to meet the needs of various natural complex scenes;
3. We collect relevant datasets using our equipment and make the datasets available for other researchers to use.

2. Method

Here, we first introduce the block diagram of our system and then introduce our SLAM subsystem and target-tracking and localization subsystem in detail, respectively. Specifically, we first introduce our three subsystems, namely VIS, LIS, and GIS. Then, we describe how to alleviate cumulative drift using global pose graph optimization. Finally, we introduce the object tracking and localization subsystem and demonstrate how to use SLAM system results to obtain the global position of the object.

2.1. The Overview of Our System

An overview of our system is shown in Figure 1, which includes a multi-sensor fusion SLAM system and an object tracking and location subsystem. The SLAM system is divided into the following three parts: data preprocessing, three internal subsystems running in parallel, and the final pose-graph optimization. The data preprocessing step preprocesses the input image, IMU, and LiDAR data, including image feature extraction, IMU pre-integration, and LiDAR plane-feature extraction. Then, it will send the results to the three subsystems, i.e., VIS, LIS, and GIS. There is an interaction between the VIS and LIS subsystems. That is, they both provide each other with the current estimated state, which can improve the accuracy and robustness of the whole system. Specifically, for VIS, we refer to the practice of sliding-window-based nonlinear optimization in VINS-Mono [6]. Since the depth of visual feature points of VIS usually has a large uncertainty, inspired by [12], we register the LiDAR point cloud to the image to assist image depth extraction, which significantly improves the accuracy of VIS for feature point depth estimation. For LIS, the large number of LiDAR point clouds leads to significant challenges in the computing performance, so we refer to the approach of ES-IEKF in FAST-LIO2 [10] and use the fast Kalman filter algorithm to accelerate the calculation. For GIS, we use the IMU data for state propagation and the GPS observation data to correct the IMU results to obtain a high-frequency GPS signal equal to the IMU frequency. Finally, we fuse the results of VIS, LIS, and GIS for pose graph optimization to correct the cumulative drift.

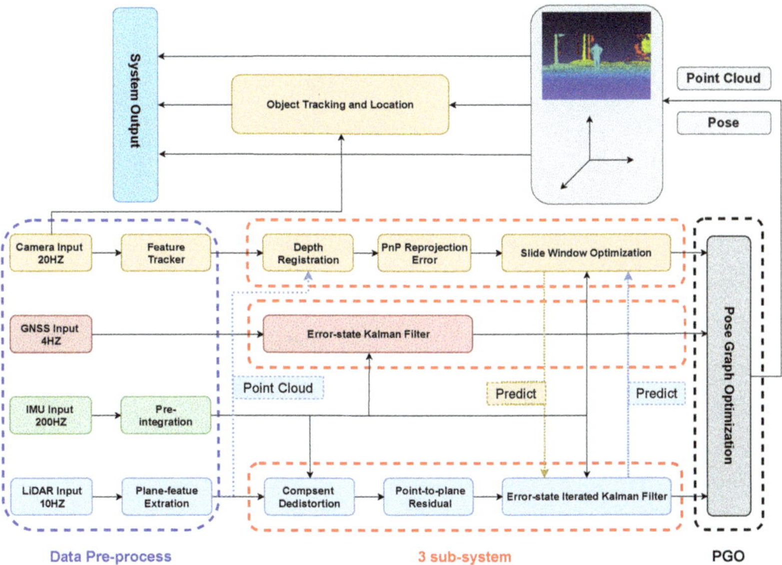

Figure 1. The overview of our system.

Target tracking and localization rely on local point cloud maps and poses of keyframes provided by the SLAM system. First, we detect targets on the image, and track them between consecutive frames using a Kalman filter. Subsequently, we filter the point clouds that fall in the detection frame based on the external parameters of the camera and LiDAR, using Euclidean clustering to filter the portion of the closest target as the target point cloud. Laser points may not occupy the ground target within the camera's field of view because the laser point cloud becomes sparse with increasing distance. Therefore, we consider using a local point cloud map instead of a single frame of laser points as our input to compensate for the sparsity of the laser point cloud. Finally, the real-time global position of the tracked target is calculated based on the key poses.

2.2. Visual Inertial Subsystem

The pipeline of VIS is similar to VINS-Mono [6], and the system block diagram is shown in Figure 1. For VIS, we define the world coordinate system as $\{W\}$, and the state variables of the IMU coordinate system are represented in $\{W\}$ as

$$\mathbf{x}_i = \left[\begin{array}{ccccc} \mathbf{R}^W_{b_i} & \mathbf{p}^W_{b_i} & \mathbf{v}^W_{b_i} & \mathbf{b}_{\omega_i} & \mathbf{b}_{a_i} \end{array}\right]^T, \tag{1}$$

where $\mathbf{R}^W_{b_i} \in SO(3)$ is the rotation matrix, $\mathbf{p}^W_{b_i} \in \mathbb{R}^3$ is the position, $\mathbf{v}^W_{b_i}$ is the velocity, $\mathbf{b}_{\omega_i}$ and $\mathbf{b}_{a_i}$ are the IMU biases. The IMU motion model is as follows:

$$
\begin{aligned}
\mathbf{p}^W_{b_{k+1}} &= \mathbf{p}^W_{b_k} + \mathbf{v}^W_{b_k}\Delta t_k \\
&\quad + \iint_{t\in[t_k,t_{k+1}]}\left(\mathbf{R}^W_t(\mathbf{a}_t - \mathbf{b}_{a_t} - \mathbf{n}_a) - \mathbf{g}^W\right)dt^2, \\
\mathbf{v}^W_{b_{k+1}} &= \mathbf{v}^W_{b_k} + \int_{t\in[t_k,t_{k+1}]}\left(\mathbf{R}^W_t(\mathbf{a}_t - \mathbf{b}_{a_t} - \mathbf{n}_a) - \mathbf{g}^W\right)dt, \\
\mathbf{q}^W_{b_{k+1}} &= \mathbf{q}^W_{b_k} \otimes \int_{t\in[t_k,t_{k+1}]}\frac{1}{2}\Omega(\omega_t - \mathbf{b}_{\omega_t} - \mathbf{n}_\omega)\mathbf{q}^{b_k}_t dt,
\end{aligned}
\tag{2}
$$

where

$$\Omega(\omega) = \left[\begin{array}{cc} -\lfloor\omega\rfloor_\times & \omega \\ -\omega^T & 0 \end{array}\right], \lfloor\omega\rfloor_\times = \left[\begin{array}{ccc} 0 & -\omega_z & \omega_y \\ \omega_z & 0 & -\omega_x \\ -\omega_y & \omega_x & 0 \end{array}\right],$$

$\mathbf{q}^W_{b_k}$ is the quaternion represent of $\mathbf{R}^W_t$, $\mathbf{a}_t$ is the acceleration reading from IMU, $\mathbf{n}_a$ is the acceleration noise of IMU, ω_t is the angular velocity reading from IMU, $\mathbf{n}_\omega$ is the angular velocity noise of IMU, Δt_k is the duration between the time interval $[t_k, t_{k+1}]$.

For the input image, VIS first detects the Fast-corner [23] and then uses the KLT algorithm [24] for optical flow tracking. Since the inverse depth of visual features optimized by VIS has great uncertainty, we accumulate the LiDAR point clouds of recent frames and then project them onto the image to assist depth estimation. Specifically, as shown in Figure 2, we use the transformation between LiDAR and camera coordinate system $\mathbf{T}^C_L = [\mathbf{R}^C_L \mid \mathbf{p}^C_L]$ to project the LiDAR point cloud onto the camera image. Then we find the nearest three projected LiDAR points on the image plane for a visual feature by searching a two-dimensional K-D tree. At last, we use these three points to fit a plane and back-project the visual feature onto this plane as its 3D point, which is shown in Figure 3. It can be seen that for most visual feature points, we can accurately estimate their corresponding 3D points. It is very beneficial in improving the accuracy of the VIS.

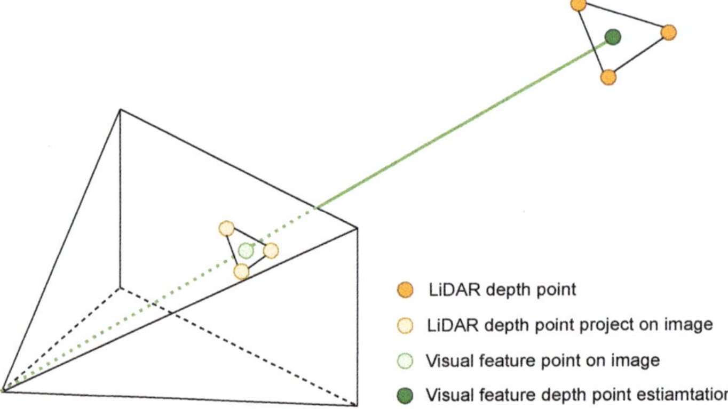

Figure 2. Visual feature depth registratoin with LiDAR points.

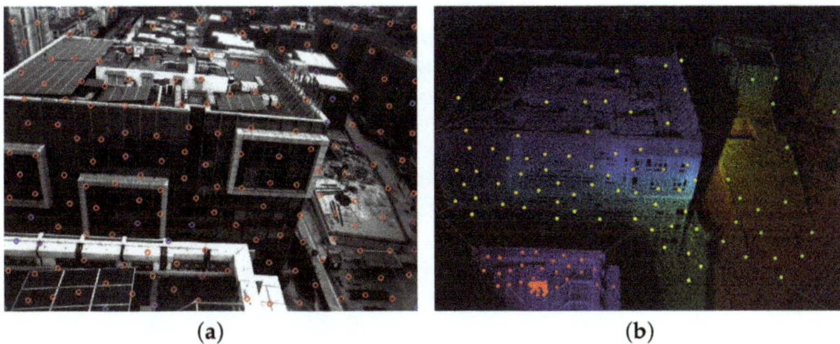

(a)   (b)

Figure 3. (**a**) The visual features on the image. (**b**) The estimated 3D points of visual feature. It's clear that lots of visual features can get accurate estimates of their 3D points.

In the back-end sliding window optimization, the camera pose and the depth of feature points will be optimized as state variables. Unlike the almost completely independent design of LIS and VIS in [12], we use the current state value x_L predicted by LIS as the initial state of VIS. The factor is added to the back-end optimization process of VIS, as shown in Figure 4. It is well known that the robustness and accuracy of LIS are higher than VIS, so this design can improve the performance of VIS, thereby making a more accurate estimate of the initial pose for the next LIS.

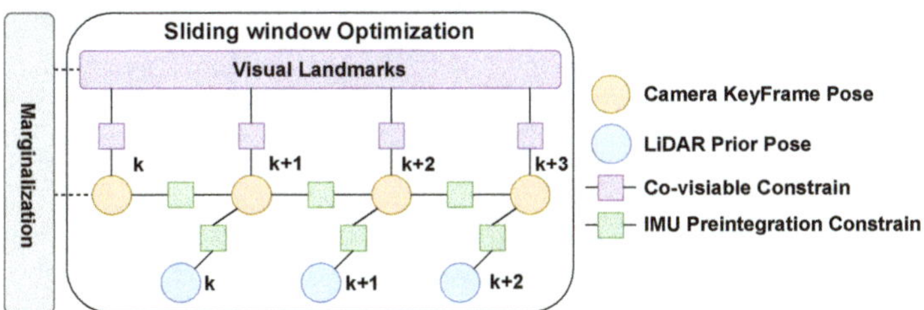

Figure 4. The factor graph of our visual inertial subsystem.

2.3. LiDAR Inertial Subsystem

Our LIS is modified from [10]. Specifically, as shown in Figure 1, we use Iterate Kalman Filter based on error-state to fuse IMU and LiDAR observations. Benefiting from the improvement of Kalman Gain K in [10], this algorithm can run in real-time without causing excessive computational burden with the increase in LiDAR observation points. The iteratively optimized Kalman filter algorithm has been proved by [25] to have the same results as the least squares algorithm using Gauss-Newton, so our LIS also guarantees the accuracy of the algorithm.

When receiving a scan from LiDAR, we first extract the plane feature points and then use the poses obtained from inertial integration to remove the motion distortion of the point cloud. We use the IMU state propagation Equation (2) to obtain an up-to-date estimate of the current LiDAR pose. However, unlike [10], thanks to the existence of our VIS, we can continue to use IMU data to estimate the current LiDAR pose based on the latest pose estimated by VIS. This method can improve the accuracy of our LIS.

After obtaining the pose estimation of current scan, we need to calculate the distance from the extracted plane feature points to the fitted plane, which is same as in [10]. However, in practical applications, the LIS has a significant drift in height, i.e., the z axis. So we add a ground constraint to solve this problem and can flexibly choose whether to use this constraint for different scenes. Our ground detection algorithm is simple but effective.

Expressly, we assume that the ground is an almost horizontal plane, which is almost always satisfied indoors and holds outdoors in the vast majority of cases. Then we accumulate the last few frames of LiDAR point clouds in the LiDAR coordinate system $\{L\}$. We assume that the distance h between LiDAR installation height and the ground is unchanged. For cars and handheld devices, this assumption is generally valid. Then we filter out all point clouds whose height is in $[h - \delta, h + \delta]$, and ground points are almost included. As shown in Figure 5 shows the ground point cloud detected by our algorithm. It can be seen that our algorithm can detect the ground well.

(a)   (b)

Figure 5. (**a**) The visual image view. (**b**) The ground point clouds(white) detected by our LiDAR inertial subsystem. It's clear that the ground point clouds can be segmented successfully.

Then we use the RANSAC algorithm to fit ground plane in these point clouds and get the equation of the ground equation in $\{L\}$. We can get

$$(\mathbf{x}_d^L - \mathbf{p}_d^L) \cdot \mathbf{n}_d^L = 0, \tag{3}$$

where $\mathbf{x}_d^L$ is point on ground plane, $\mathbf{p}_d^L$ is point on the plane and $\mathbf{n}_d^L$ is the plane normal vector in the $\{L\}$. Since world coordinate system $\{W\}$ is aligned with gravity $\mathbf{g} = [0, 0, -g]^T$, we can easily get the actual ground equation in our world coordinates as

$$(\mathbf{x}_t^W - \mathbf{p}_t^W) \cdot \mathbf{n}_t^W = 0, \tag{4}$$

where $\mathbf{p}_t^W = [0, 0, -h_0]$ and $\mathbf{n}_t^W = [0, 0, 1]^T$, and h_0 is the distance from the origin of our $\{W\}$ coordinate system to the ground.

In the same way, we use the current LiDAR pose $\mathbf{T}_L^W$ to convert this actual ground equation into the current frame LiDAR coordinate system $\{L\}$. We can get

$$(\mathbf{x}_t^L - \mathbf{p}_t^L) \cdot \mathbf{n}_t^L = 0. \tag{5}$$

Next, we define the detected plane $[\mathbf{n}_d^L; \mathbf{p}_d^L]$, the real plane $[\mathbf{n}_t^L; \mathbf{p}_t^L]$, and the error between them is added to the optimization of LIS to alleviate the z axis drift. First we use a rotation matrix $\mathbf{R} \in SO(3)$ to rotate $\mathbf{n}_t^L$ to x axis, that is

$$\begin{bmatrix} 1 & 0 & 0 \end{bmatrix}^\top = \mathbf{R}\mathbf{n}_t^L. \tag{6}$$

Then we use this rotation matrix to rotate $\mathbf{n}_d^L$, that is

$$\begin{bmatrix} x_{d'}^L & y_{d'}^L & z_{d'}^L \end{bmatrix}^\top = \mathbf{R}\mathbf{n}_d^L. \tag{7}$$

We define the variables to measure the orientation error between the detected ground equation and the actual ground equation as

$$\alpha = \arctan \frac{y^L_{d'}}{x^L_{d'}}, \quad \beta = \arctan \frac{z^L_{d'}}{\sqrt{x^L_{d'}{}^2 + y^L_{d'}{}^2}}. \tag{8}$$

Define the variables to measure the translation error of these two planes as

$$\gamma = \mathbf{p}^L_t \cdot \mathbf{n}^L_t - \mathbf{p}^L_d \cdot \mathbf{n}^L_d. \tag{9}$$

We define $\mathbf{e} = [\alpha, \beta, \gamma]^T$ as the final ground residual, which can well constrain the error of our system in the directions of *roll*, *pitch*, and *z*. This improvement on the final LIS can be seen in our later experiments.

2.4. GPS Inertial Subsystem

We add a GPS-inertial subsystem to better prepare for back-end pose graph optimization. In this subsystem, we use the Error-state Kalman Filter to integrate the state obtained by IMU and the state of GPS observations. This fusion can obtain a high-frequency pose output at IMU frequency. The reason for this design is that the frequency of GPS data is relatively low, and it cannot accurately time matched with the output of the VIS or LIS [9,15] both use the assumption of uniform motion to interpolate GPS data. However, in the case of high-speed and non-uniform motion, this assumption will cause the interpolation results to contain large noise, which reduces the accuracy of back-end pose graph optimization. Specifically, we use Equation (2) to predict the IMU state and define the error of the state $\mathbf{x}$ prediction as

$$\delta\mathbf{x} = \begin{bmatrix} \delta\mathbf{R}^W_b & \delta\mathbf{p}^W_b & \delta\mathbf{v}^W_b & \delta\mathbf{b_\omega} & \delta\mathbf{b_a} \end{bmatrix}^\top. \tag{10}$$

For GPS data, it is defined in WGS84 coordinate system. Same as [15], we first convert the data to ENU coordinate system as our state. Assuming that the position of our GPS sensor in the IMU coordinate system $\{b\}$ is $\mathbf{p}^b_{GPS}$, then we can get GPS data to observe the origin of IMU coordinate system $\mathbf{p}^W_{GPS}$ as

$$\mathbf{p}^W_{GPS} = \mathbf{p}^W_b + \mathbf{R}^W_b \mathbf{p}^b_{GPS}. \tag{11}$$

Using Equation (11) we can get the Jacobian matrix of GPS observations for the error state $\delta\mathbf{x}$ as

$$\mathbf{x}_{GPS} = \begin{bmatrix} \mathbf{I} & \mathbf{0} & -\mathbf{R}^W_b \lfloor \mathbf{p}^b_{GPS} \rfloor_\times & \mathbf{0} & \mathbf{0} \end{bmatrix}^\top. \tag{12}$$

Referring to the error-state Kalman filter equation in [26], we can get the result of the fusion of GPS and IMU, which is more accurate than the interpolation of GPS data using the assumption of uniform motion.

2.5. Pose Graph Optimization

After all three subsystems complete their estimation tasks, their results are sent to the final pose graph optimization system for processing. In the pose graph optimization system, we select keyframes for the input of VIS and LIS and use iSAM2 [27] to optimize the pose graph. Precisely, we will count the pose changes between the latest keyframes in the relative pose map of the current input frame. If the rotation or translation part of the pose transformation exceeds the threshold we set, then we will use it as an optimized keyframe. Thanks to our GIS, it uses GPS and IMU to perform Error-state Kalman Filter to get high-frequency GPS observations, which makes a GPS observation constraint almost available for each keyframe. However, the GPS signal usually has a large error when occluded, which reflected in our GIS system is the output covariance P_{GPS} is relatively large. We filter the results of $P_{GPS} < \theta_P$ to add to pose graph. As shown in Figure 6, it is the block diagram of pose graph optimization system.

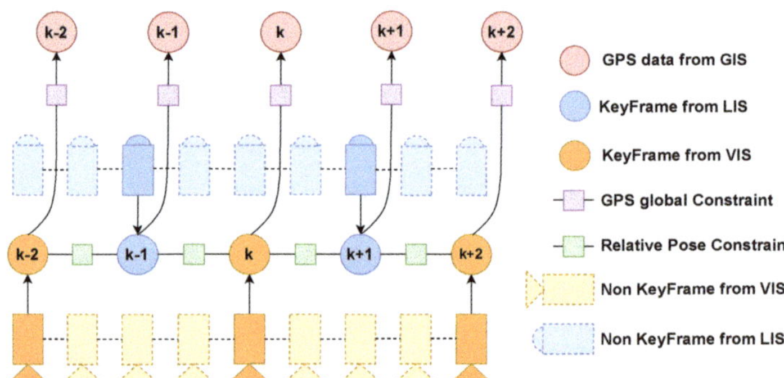

Figure 6. The block diagram of pose graph optimization.

2.6. Targets Detection and Tracking

We use YOLOv4 to detect the targets and track them with the optical flow method and Kalman filter. We first use the optical flow method to eliminate the deviation of the pixel coordinates caused by the camera movement of targets and then use the Kalman filter to predict their position.

2.6.1. L-K Optical Flow

First, the L-K optical flow assumes that the gray value of the same point in space is constant across images; then, the weighted least squares method is used to estimate the optical flow field, where the grayscale value of the point $a = (x, y)$ at time t is assumed to be $I = (x, y, t)$, and the optical flow constraint equation can be derived based on the following assumptions:

$$\nabla I \cdot V_a + I_t = 0, \tag{13}$$

where $\nabla I = (I_x, I_y)$ denotes the gradient of the image at point a; and $V_a = (u, v)$ is the optical flow at point a. Assuming that the optical flow is the same at each point in a local neighborhood centered at point a, search for the displacement that minimizes the matching error in this block, i.e., define Equation (14) for this neighborhood, and minimize its function value as follows:

$$F(x, y) = \sum_{(x,y) \in \Omega} W^2(x, y) \ [\nabla I \cdot V_a + I_t], \tag{14}$$

where Ω denotes the local neighborhood of point a and $W(x, y)$ denotes the weight function. The optimal solution of equation Equation (14) is obtained as follows:

$$\begin{aligned}
A &= [\nabla I(x_1), \nabla I(x_2), \cdots, \nabla I(x_n)], \\
W &= \text{diag}[W(x_1), W(x_2), \cdots, W(x_n)], \\
b &= -[I_t(x_1), I_t(x_2), \cdots, I_t(x_n)].
\end{aligned} \tag{15}$$

The final equation can be solved:

$$V = \left[A^\mathsf{T} W^2 A \right]^{-1} A^\mathsf{T} W^2 b. \tag{16}$$

The simple L-K optical flow method cannot manage a situation where the UAV is moving quickly at a high altitude. Moreover, it will generate significant computational errors due to the large motion, which will not only affect the algorithm's accuracy but also reduce the overall computing speed. In this paper, we employ the pyramid-based L-K optical flow method, whose principle is described as follows: First, the optical flow and affine transformation matrices are calculated for the image of the highest layer. The result of the calculation of the previous layer initializes the calculation of next layer. The optical flow and affine transformation matrices are calculated based on the initialization. This process is repeated until the original image layer is reached. The final result will be computed depending on this coarse-to-fine filtering process.

2.6.2. Kalman Filter Update Objective

In the case of a video that needs to be tracked, the state vector can be expressed as follows:

$$\mathbf{X} = \begin{bmatrix} \mathbf{c}_x & \mathbf{c}_y & \mathbf{w} & \mathbf{h} & \mathbf{v}_x & \mathbf{v}_y & \mathbf{v}_w & \mathbf{v}_h \end{bmatrix}, \tag{17}$$

where $\mathbf{c}_x, \mathbf{c}_y$ are centers, $\mathbf{w}, \mathbf{h}$ are the width and height of the bounding box, and $\mathbf{v}_x, \mathbf{v}_y, \mathbf{v}_w, \mathbf{v}_h$ are the speed of their change. Note that $\mathbf{c}_x, \mathbf{c}_y$ are corrected in Section 2.6.1. We can predict $\mathbf{X}_t$ based on $\mathbf{X}_{t-1}$:

$$\mathbf{x}' = \mathbf{Fx} + \boldsymbol{\mu}, \tag{18}$$

$$\mathbf{P}' = \mathbf{FPF}^\top + \mathbf{Q}, \tag{19}$$

where $\mathbf{F}$ is called the state transfer matrix, $\mathbf{P}$ is the covariance of tracking at the moment $t-1$, and $\mathbf{Q}$ is the noise matrix of the system. The observed and predicted values are considered to apply to the same target if they satisfy the following two conditions:

- The current detection frame is expanded with enough overlap area with the prediction frame;
- The Reid score is higher than a specific value.

Finally the tracker is updated with predicted ($\mathbf{x}'$) and observed values, as follows ($\mathbf{y}$):

$$\mathbf{y} = \mathbf{z} - \mathbf{Hx}', \tag{20}$$

$$\mathbf{S} = \mathbf{HP}'\mathbf{H}^\top + \mathbf{R}, \tag{21}$$

$$\mathbf{K} = \mathbf{P}'\mathbf{H}^\top \mathbf{S}^{-1}, \tag{22}$$

$$\mathbf{x} = \mathbf{x}' + \mathbf{Ky}, \tag{23}$$

$$\mathbf{P} = (\mathbf{I} - \mathbf{KH})\mathbf{P}', \tag{24}$$

where the observation matrix $\mathbf{Z} = \begin{bmatrix} \mathbf{c}_x & \mathbf{c}_y & \mathbf{w} & \mathbf{h} \end{bmatrix}$.

2.6.3. LiDAR Vision Fusion Targeting

To calculate the point cloud falling in the bounding box, we first calculate the pixel coordinates of the laser point in the image as follows:

$$\begin{bmatrix} \mathbf{c}_i^{cam} \\ 1 \end{bmatrix} = \mathbf{T}_b^c (\mathbf{T}_b^w)^\top \begin{bmatrix} \mathbf{c}_i^{map} \\ 1 \end{bmatrix}, (i = 1,2,3,\ldots,N_A), \tag{25}$$

$$\mathbf{p}_i^{cam} = \begin{bmatrix} u_i \\ v_i \\ 1 \end{bmatrix} = \frac{\mathbf{I}^{cam}}{w_i'} \mathbf{c}_i^{cam} = \frac{1}{w_i'} \begin{bmatrix} u_i' \\ v_i' \\ w_i' \end{bmatrix}, (i = 1,2,3,\ldots,N_A), \tag{26}$$

where $\mathbf{c}_i^{map} = [x_i, y_i, z_i]^\top$ is the point in the local point cloud map and N_A is the amount of point clouds, $\mathbf{T}_b^w = [\mathbf{R}_b^w \mid \mathbf{p}_b^w]$, $\mathbf{T}_b^c = [\mathbf{R}_b^c \mid \mathbf{p}_b^c]$, $\mathbf{p}_i^{cam}$ is the pixel coordinate of the point cloud, and $\mathbf{I}^{cam} \in \mathbb{R}^3 \times 3$ is the intrinsic matrix of the camera.

Suppose there are n trackers in the current image. We filter the point clouds $\mathbf{C}^{cam} = \{\mathbf{c}_i^{cam} | i = 1,2,3,\ldots,N_A\}, \mathbf{C}^{map} = \{\mathbf{c}_i^{map} | i = 1,2,3,\ldots,N_A\}$ in the occupancy detection frame and use Euclidean clustering to classify the points in the map point cloud $(\mathbf{C}^{map})'$ into n classes. Moreover, the center of the point cloud $\mathbf{O} = \{\mathbf{o}_i | i = 1,2,3,\ldots,n\}$ is the target object's location. The largest rectangle that can wrap the point cloud represents the outline of the target.

3. Physical Experiment Analysis

Considering that there is no open dataset that can satisfy all the sensors used by our algorithm simultaneously, we built our hardware equipment and collected a large number of scene-rich datasets to verify our algorithm. First, we will introduce our hardware equipment and the collected datasets. Then, based on the collected dataset, we experimentally verify the improvements made in our system relative to [6,9]. Specifically, our experiments included the improvement of VIS by registering feature depth with LiDAR and improvements related to leading ground constraints into LIS. Then, we tested the improvement of the positioning and mapping accuracy of the SLAM system by adding a GPS to pose graph optimization on datasets of different scales.

We set up two experiments for the target localization system on the ground and in the air, respectively. The ground experiment mainly verifies the Ray-vision fusion's relative localization effect without providing the global localization error. After that, we conducted aerial experiments while measuring the global positioning error at different distances based on our SLAM. All experiments were performed on the same system with an Intel® Core™ i7-9700 CPU @ 3.00GHz × 8 and Nvidia GTX 1080ti.

3.1. Hardware and Dataset of Our System

The hardware of our system is shown in Figure 7, which includes a global shutter camera, a LiVox AVIA LiDAR (FoV of 70.4° × 77.2°), a GNSS-INS module, a power supply unit, and an onboard computation platform (equipped with an Intel i5-8400 CPU and 16 GB RAM).

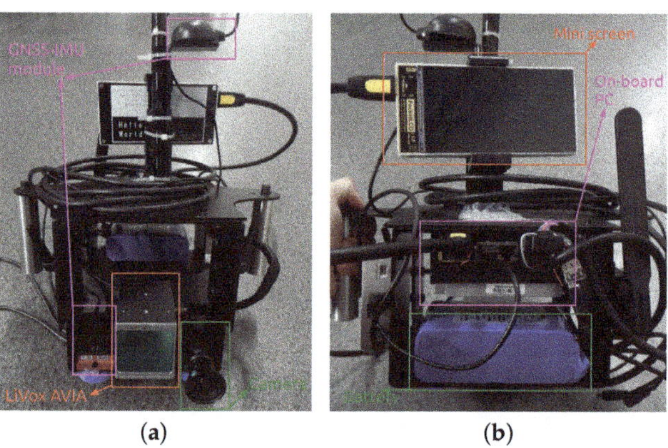

Figure 7. (**a**) The front view of our hardware. (**b**) The back view of our hardware. The total weight of our device is below 3 kg.

We collected various datasets with rich scenes. Specifically, we used two large-scale datasets collected by drones at an altitude of about 100 m, which we call HZ-odom and HZ-map. In the ground scene, we fixed the device to the electric bicycle and collected several datasets, including two large-scale datasets, which we named ZJG-gym and ZJG-nsh. Three medium-sized datasets, which we name ZJG-lib, YQ-odom, and YQ-map. We also collected two hand-held datasets of small-scale scenarios on the ground and named them CSC-build and CSC-road. The specific information of each dataset is in Table 1, which includes the duration of the dataset, the length of the trajectory, whether they include a return to the origin, and the difficulty level.

Table 1. Dataset detail.

Dataset	Duration (s)	Length (m)	Loop Closure	Difficulty Level
HZ-odom	280	2380	No	Difficult
HZ-map	367	3068	No	Difficult
ZJG-gym	612	2761	Yes	Difficult
ZJG-nsh	576	1668	Yes	Difficult
ZJG-lib	239	866	Yes	Medium
YQ-odom	306	1718	Yes	Medium
YQ-map	256	1298	Yes	Medium
CSC-road	80	117	Yes	Easy
CSC-build	73	86	Yes	Easy

3.2. Feature Depth Registration of VIS

In this experiment, we focus on the improvement achieved by including the depth registration of visual feature points using LiDAR point cloud in our VIS system. Since our VIS system is adapted from VINS-Mono [6], we mainly compare the results of our VIS system for depth registration of visual feature points with [6]. During the experiment, our LIS subsystem only runs the point cloud preprocessing part. It does not use its running results to add priors to VIS. The block diagram of our VIS in the experiment is shown in Figure 8.

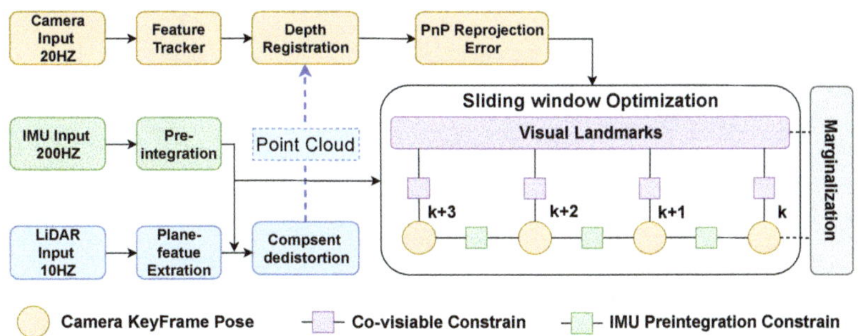

Figure 8. The block diagram of our VIS experiment.

We conducted experiments on YQ-map ground dataset. In this dataset, the vehicle runs at a constant speed most of the time; therefore, the IMU is close to degenerating. Moreover, the vehicle has many 90° turns, which means the visual constraints of the VIO may easily fail, and this feature can lead to scale drift problems. For fairness of the experiment, we turned off the loop closure detection thread of VINS-Mono and only compared the accuracy of the odometry. Furthermore, we set VINS-Mono and our VIS system to be identical in terms of front-end feature extraction, back-end sliding-window keyframes, and optimization time. Since the YQ-map dataset returns to the origin, we use the distance from the endpoint to the start point to judge the accuracy. The result is shown in Figure 9. Due to the depth registration of visual feature points, we can see that our system has better scale consistency and higher accuracy than VINS-Mono. This result is easy to explain: our VIS outperforms VINS-Mono due to the extra scale gained by adding LiDAR point clouds to the depth registration of visual feature points.

In addition, to test the improvement of the absolute accuracy of VIS by performing depth registration, we also conduct experiments on large-scale dataset ZJG-gym and use GPS trajectories as ground truth. The comparison results of our VIS and VINS-Mono are shown in Figure 10. We can see that our VIS and GPS trajectories are in better alignment. We use the root mean square error (RMSE) results to measure the result accuracy. The RMSE of our VIS is 5.03 m, while VINS-Mono is 7.92 m.

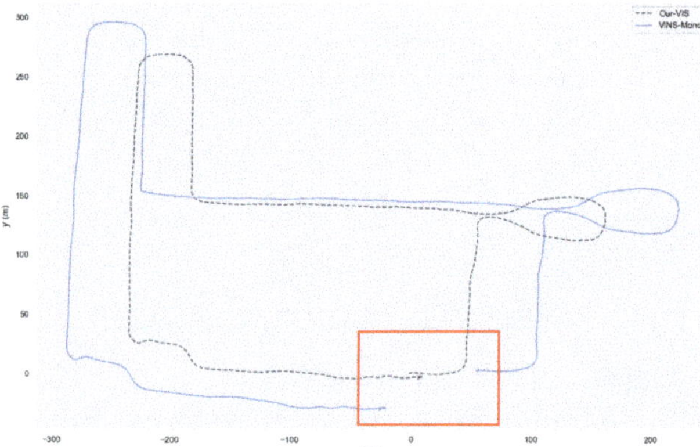

Figure 9. The result of our VIS and VINS-Mono on YQ-map dataset. It goes back to origin.

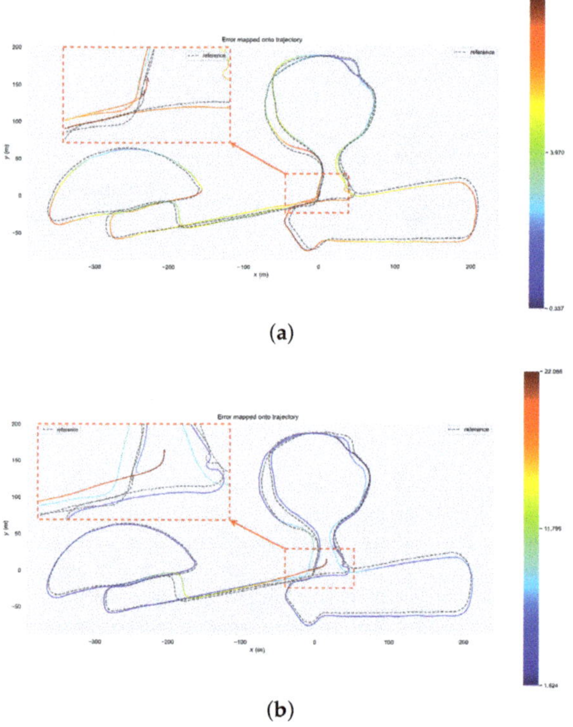

(a)

(b)

Figure 10. (**a**) The trajectory alignment of our VIS and GPS on ZJG-gym dataset. (**b**) The trajectory alignment of VINS-Mono and GPS on ZJG-gym dataset. It is clear that our VIS has a better result.

3.3. Ground Constraint of Our LIS

In this experiment, we focus on the improvement by adding ground constraints to our LIS subsystem. Since our LIS system is adapted from FAST-LIO2 [10], we mainly compare the results of our LIS system with ground constraints and [10]. Specifically, in the experiments, we run our LIS alone without VIS's prediction.

We conducted our experiments on ZJG-lib dataset, which has many horizontal grounds. Specifically, our LIS system uses the same parameter configuration as FAST-LIO2, including the point cloud downsampling density, the number of iterative Kalman filtering, etc. Then in the experiment, we focus on the *z* axis drift. The result is shown in Figure 11. From the figure, we can see that due to the ground constraints we added, our LIS system has almost no drift in the height direction, while FAST-LIO2 shows a significant drift in height. This result is easy to explain. For ZJG-lib dataset, the LiDAR installation location is close to the ground with a height of 0.8 m. Therefore, the incident angle of the LiDAR scanning distant ground points is small, which reduces the accuracy of the point cloud scanned by LiDAR and increases the drift in altitude of FAST-LIO2.

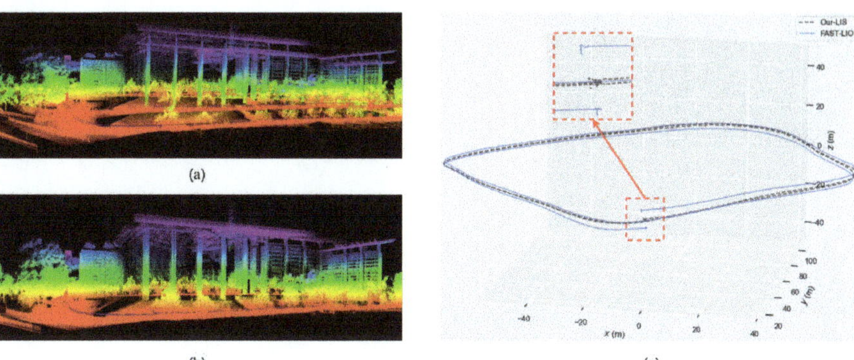

Figure 11. (**a**) The map result of FAST-LIO2 on ZJG-lib dataset. (**b**) The map result of our LIS with ground constraint on ZJG-lib dataset. (**c**) The path result of FAST-LIO2 and our LIS. It is clear that due to the ground constraint, our LIS has less drift than FAST-LIO2.

3.4. Pose Graph Optimization of Our System

In this experiment, we focus on the improvement of our entire SLAM system due to the addition of global pose graph optimization. To demonstrate the cumulative trajectory drift suppression that occurs when incorporating a GPS, we conducted experiments on large-scale datasets in the air and on the ground. Specifically, we used multiple large-scale datasets to examine the improvement of our system by fusing GPS data.

First, we conduct experiments on an aerial dataset with HZ-map dataset, in which the drone carries equipment to collect data at an altitude of 100 m. Because of the poor weather conditions when collecting the data, the aircraft in the air is highly unstable, so this dataset presents a considerable limitation in terms of the accuracy and robustness of the SLAM system. Here, we compare the localization and mapping of our system with VINS-Mono and FAST-LIO2 systems, and the results are shown in Figure 12. Since our system fuses GPS data, there is better consistency in positioning and mapping results in large-scale scenes. However, due to the turbulence of the drone at a high altitude in this dataset, VINS-Mono fails and does not provide meaningful results.

In order to better test the accuracy and robustness of our algorithm, we also conducted experiments on many other datasets and used the GPS trajectory as the ground truth. We show the root mean square error (RMSE) results in Table 2 and some trajectory results in Figure 13. The compared algorithms are the separate VIO system VINS-Mono [6], the separate LIO system FAST-LIO2 [10], and the state-of-the-art LIVO systems R2LIVE [13] and R3LIVE [28], which are most similar to our system. It is clear that our system achieves the best accuracy and the most robust performance. Interestingly, after R2LIVE [13] and R3LIVE [28] are integrated with cameras, some data sequences have a lower precision than FAST-LIO2 [10]. This is expected because incorporating visual information in situations that are not conducive to camera work may reduce accuracy. Therefore, visual information is usually incorporated in LIO systems mainly to increase the robustness of the whole system. In addition, R2LIVE [13] and R3LIVE [28] failed on both HZ-map and HZ-Odom datasets, which are run in harsh high-altitude environments. Their two subsystems, VIO and LIO, are tightly coupled. Once a subsystem state is incorrectly estimated, it will have a devastating impact on the entire system. Because our VIO and LIO subsystems are loosely coupled, the collapse of one subsystem will not affect the regular operation of the whole system.

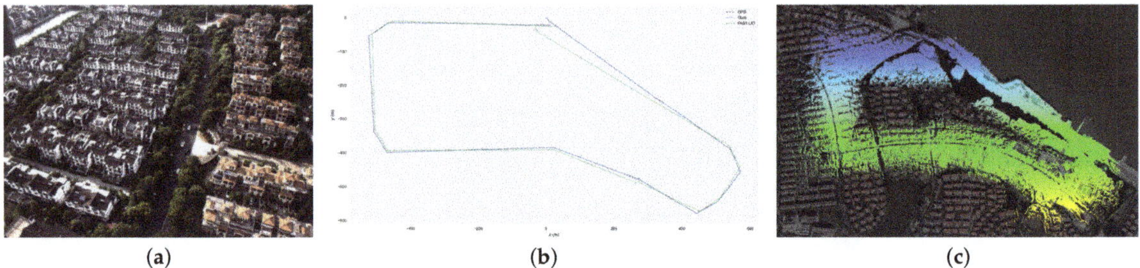

(a) (b) (c)

Figure 12. (**a**) The image view of our HZ-map dataset. (**b**) The trajectory of GNSS, FAST-LIO2 and our system. VINS-Mono failed due to the aggressive motion. (**c**) The map built by our system aligned with Google Earth. As a result of the GNSS fusion in our system, we obtain a highly accurate result.

Table 2. RMSE translation error w.r.t GPS.

Algorithm	HZ-Map	HZ-Odom	YQ-Map	YQ-Odom	ZJG-Gym	ZJG-Nsh	ZJG-Lib	CSC-Road	CSC-Build
VINS-Mono [6]	Failed	Failed	50.23	2.34	7.92	6.38	39.00	0.57	0.66
FAST-LIO2 [10]	15.20	0.96	1.80	1.34	3.30	1.88	3.22	0.35	0.43
R2LIVE [13]	Failed	Failed	1.70	12.18	3.36	1.95	3.46	0.26	0.37
R3LIVE [28]	Failed	Failed	1.68	1.37	2.95	2.04	3.78	0.28	0.34
Ours	**0.72**	**0.55**	**0.75**	**0.75**	**2.94**	**0.97**	**1.60**	**0.24**	**0.29**

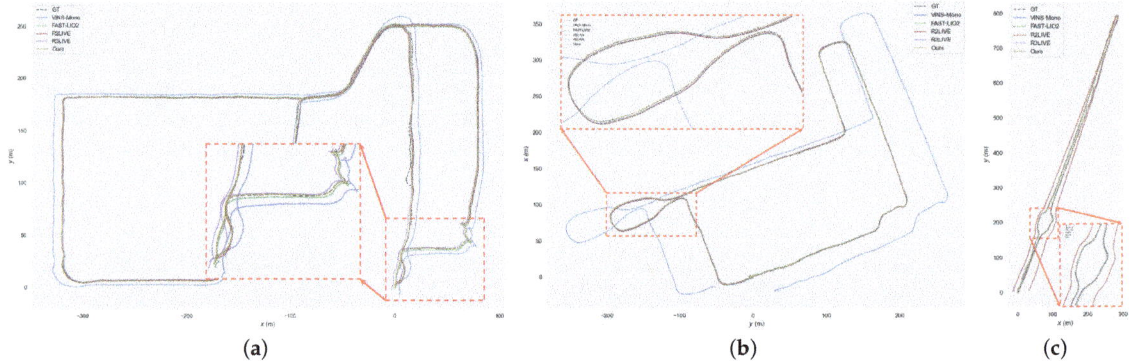

(a) (b) (c)

Figure 13. (**a**) The trajectory results on ZJG-nsh dataset. (**b**) The trajectory results on YQ-map dataset. (**c**) The trajectory results on YQ-odom dataset.

3.5. Robustness Evaluation of SLAM System

First, we evaluate the robustness of our system when subject to severe motion. HZ-map is a dataset of drones flying in the air. During automatic flight, the plane experiences sudden stops, turns, and other actions. As shown in Figure 14, the plane has a large number of drastic pitch and yaw angle changes. It can also be seen from the previous experiments that VINS-Mono, R2LIVE, and R3LIVE all failed. Our VIS subsystem also failed. However, the other two subsystems still usually work, which showes the excellent robustness of our system.

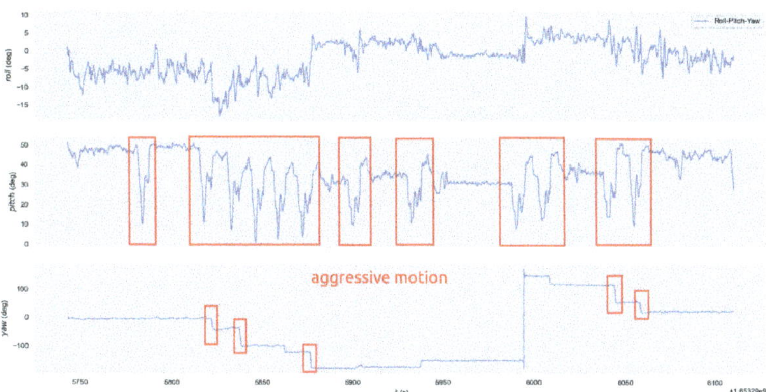

Figure 14. The Euler Angles of HZ-map dataset, which has an aggressive motion.

Then, we tested the performance of our system when the sensor failed on YQ-map dataset. As shown in Figure 15a, we selected three locations uniformly throughout the trajectory and disabled the camera, LiDAR, and GPS to detect the impact on the system. The experimental results are shown in Figure 15b. When one or two sensors fail, our system can operate normally and obtain relatively accurate trajectory results. Running only the VIO system leads to poor trajectory results only when both the GPS and LiDAR fail. This experiment proves that our system has good robustness.

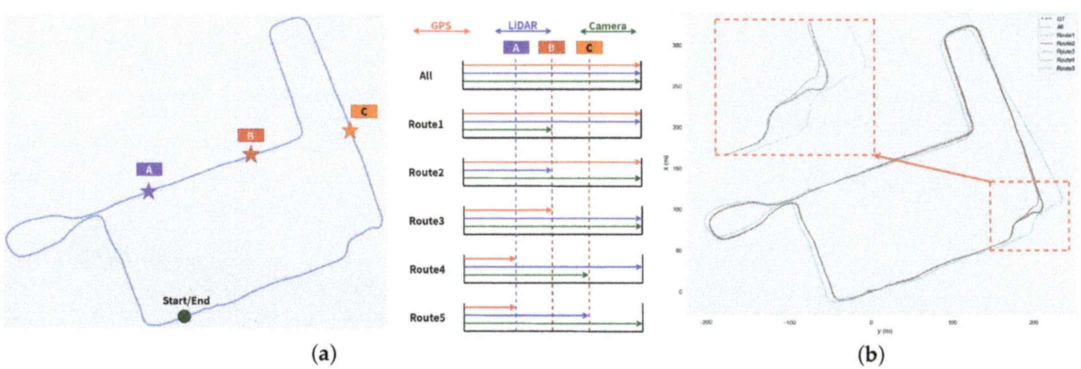

Figure 15. (**a**) Three locations throughout the trajectory to disable the camera, LiDAR, and GPS. (**b**) The trajectory result of our system when one or more sensors are closed.

3.6. LiDAR-Vision Fusion Relative Localization

One person was arranged in an open outdoor scene as the target to be located in the experiment. The person holds RTK, as shown in Figure 7, facing a dynamic target walking arbitrarily within 10–40 m of the device's field of view (as shown in Figure 16). We randomly select some locations as checkpoints and use the laser to measure the distance between the target and the sensor module to compare with the localization results and evaluate the relative localization accuracy of the algorithm. The experimental results are shown in Table 3.

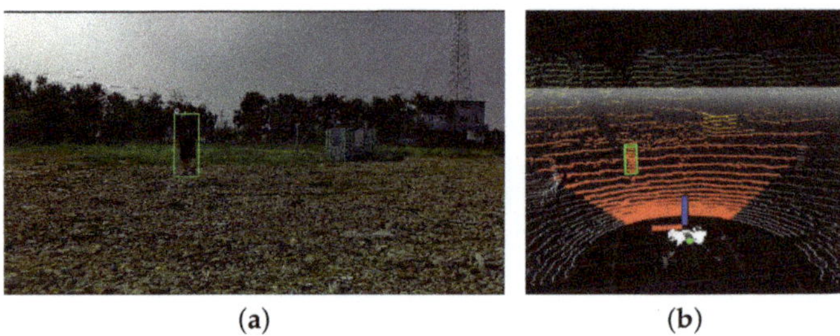

(a)   (b)

Figure 16. Two test screenshots are shown. (**a**) The results of image detection and tracking. The target is marked with a green box, and the point cloud in (**b**) is also marked with a green box to indicate the spatial position relative to the perception module.

Table 3. Relative Positioning error.

No.	x (m)	y (m)	$\sqrt{x^2+y^2}$ (m)	Truth (m)	Error (m)
1	15.19	−1.70	15.285	15.30	0.015
2	18.63	−2.43	18.788	19.20	0.412
3	24.76	3.56	25.015	25.0	−0.015
4	25.96	−3.68	26.220	26.30	0.0.8
5	28.82	−9.97	30.496	30.50	0.004
6	35.84	−1.41	35.868	35.90	0.032
7	26.47	4.33	26.821	26.80	−0.021
mean	-	-	-	-	0.073

As seen from the results in Table 3, the proposed LiDAR-Camera fusion positioning method in this paper has high measurement accuracy: the relative positioning error is provided in centimeters when the target is less than 40 m away from the measurement unit. As GPS is usually only accurate to the meter level; therefore, it is essential first to ensure that the relative positioning accuracy is as high as possible to maintain the lowest possible error when converting the object's position to the world coordinate system. The following describes the global positioning experiment for the target.

3.7. SLAM Based Global Localization

We chose to perform this experiment in the air, keeping with the system's actual application scenario. Again, we chose people as targets to evaluate our algorithm. As shown in Figure 17, our UAV flies to a distance of 10 m, 30 m, 60 m, and 90 m from the target for ground reconnaissance, while on the other hand, the target on the ground carries a GNSS receiver that moves at least ten meters along a given trajectory. At this point, the UAV is stationary or follows the target in motion, keeping the target in view. Once the separation distance exceeds 60 m, it is difficult to observe the target in the image with the naked eye. Therefore, we retrained the detection model using the Visdrone Dataset [29] and our small air-to-ground target dataset (Table 4).

Table 4. Comparison of Visdrone and our Dataset.

Image Object Detection	Scenario	Images	Categories	Avg. Labels/Categories	Resolution (m)	Occlusion Labels
Visdrone [29]	drone	10209	10	54.2 k	2000 × 1500	✓
Ours	drone	3625	2	13.1 k	1440 × 1080	✓

On the other hand, the laser point cloud at high altitudes would be sparse in terms of measuring the distance to the target. Therefore we use the local point cloud map provided by our SLAM system for the calculation instead of the single frame. In addition, our SLAM system also provides the UAV pose corresponding to the local point cloud, which can be used to calculate the global positions of the targets. It is worth mentioning that our framework locates all targets in the field of view in real time. However, to facilitate the evaluation of the results, we only count the localization error of one of the targets (Figure 18). The RTK conversion on our target path point to the coordinates under the takeoff point ENU coordinate system is used as ground truth. The distance between the localization result and its closest ground truth is used as the error to evaluate the algorithm.

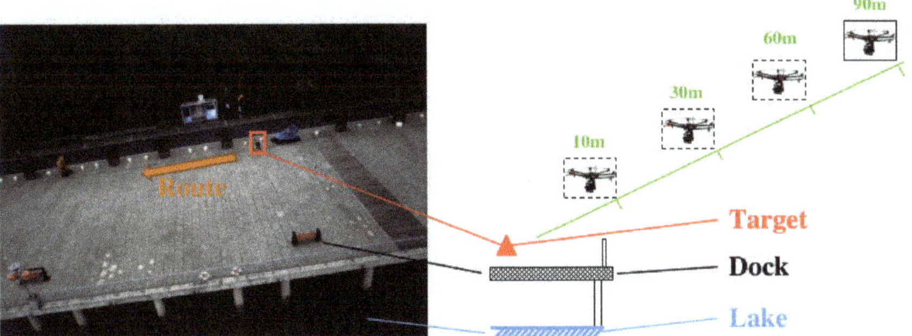

Figure 17. Schematic diagram of the aerial reconnaissance experiment. The image on the left is the actual aerial image (about 30 m apart), the red box is the target's position, and the image on the right indicates the relative position relationship between the UAV and the target.

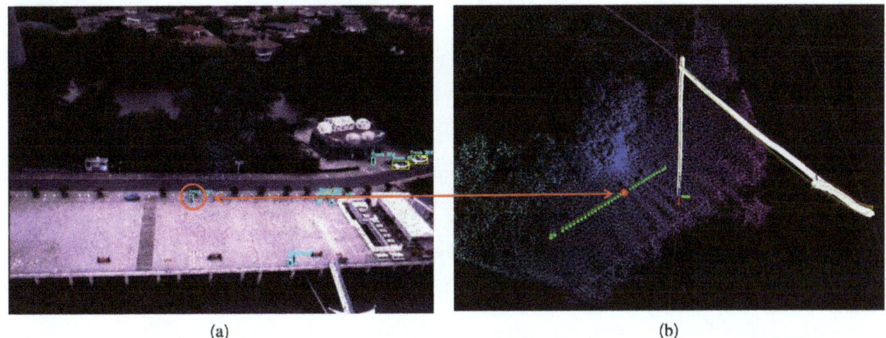

(a)   (b)

Figure 18. The display image of the tracking effect. (**a**) shows the real-time tracking result, and after a target is selected, the target is indicated by a red ball in (**b**). The green in (**b**) represents ground truth, and the white color is the aircraft's trajectory calculated by our SLAM.

We collected data using a UAV and verified our algorithm offline. The target position was inferred in 71.5 ms in one round of the experiment. The experimental results are shown in Table 5. Even at high altitudes where the sensors moved violently, our algorithm tracked the target stably and maintained high positioning accuracy. Especially at medium and long distances of 90 m, where the target occupied only a dozen pixel values, our system maintained an error of about one meter.

Table 5. Global Positioning error.

Distance (m)	Amount	Min (m)	Max (m)	Median (m)	SD (m)	MAE (m)
10 ($\pm$1)	268	0.16	1.86	0.63	0.30	0.70
30 ($\pm$1)	200	0.26	3.63	0.99	0.54	1.08
60 ($\pm$2)	216	0.07	2.19	0.71	0.35	0.74
90 ($\pm$3)	158	0.02	3.91	1.09	0.59	1.11

4. Discussion Conclusions

In the case of GPS denial, UAV self-positioning and target detection technology can play a very influential role in the military and rescue fields that require reconstructing the target area's ground scene quickly, obtaining the corresponding GPS position, and marking the category and real-time position of some critical targets.

This paper proposes a robust, versatile self-localization mapping and target-tracking localization system. Our SLAM system fuses multiple local and global sensors, including camera, LiDAR, IMU, and GPS, and thus has the advantages of high local accuracy and no global drift. Our SLAM system consists of three subsystems, VIS, LIS, and GIS. The three subsystems are tightly coupled, and the subsystems are loosely coupled through information sharing. This system architecture not only ensures the system's accuracy but also improves the system's robustness. We built our experimental equipment, collected many air and ground datasets, and conducted detailed experimental verification and analysis. The experimental results prove that our SLAM system has higher accuracy and better robustness than the current SOTA system. We also introduce a LiDAR-Camera Fusion object tracking and localization algorithm. We first used the retrained YOLOv4 to detect the target's position on the image and used LK optical flow and the Kalman filter to track the targets. We used LiDAR to recover the depth Information of the Target. Our target-tracking and localization system can effectively detect the target of interest and perform global localization of the target based on the results of the SLAM system. We conducted extensive experiments on our collected datasets, and the results show that our system performs the expected functions well. In the future, we will research real-time online tightly coupled GPS data and the high-altitude detection of small targets.

Author Contributions: Methodology, C.C. and Y.M.; data curation, L.L.; writing—original draft preparation, C.C. and Y.M.; writing—review and editing, J.L. and X.Z.; supervision, Y.L.; project administration, W.G. All authors have read and agreed to the published version of the manuscript.

Funding: This research received no external funding.

Institutional Review Board Statement: Not applicable.

Informed Consent Statement: Not applicable.

Data Availability Statement: Not applicable.

Conflicts of Interest: The authors declare no conflict of interest.

References

1. Fink, G.; Franke, M.; Lynch, A.F.; Röbenack, K.; Godbolt, B. Visual inertial SLAM: Application to unmanned aerial vehicles. *IFAC-PapersOnLine* **2017**, *50*, 1965–1970. [CrossRef]
2. Tang, J.; Xu, D.; Jia, K.; Zhang, L. Learning parallel dense correspondence from spatio-temporal descriptors for efficient and robust 4d reconstruction. In Proceedings of the IEEE/CVF Conference on Computer Vision and Pattern Recognition, Nashville, TN, USA, 20–25 June 2021; pp. 6022–6031.
3. Kyriakaki, G.; Doulamis, A.; Doulamis, N.; and Ioannides, M.; Makantasis, K.; Protopapadakis, E.; Hadjiprocopis, A.; Wenzel, K.; Fritsch, D.; Klein, M.; et al. 4D reconstruction of tangible cultural heritage objects from web-retrieved images. *Int. J. Herit. Digit. Era* **2014**, *3*, 431–451. [CrossRef]
4. Mustafa, A.; Kim, H.; Guillemaut, J.; Hilton, A. Temporally coherent 4d reconstruction of complex dynamic scenes. In Proceedings of the IEEE Conference on Computer Vision and Pattern Recognition, 2016, pp. 4660–4669.
5. Mourikis, A.I.; Roumeliotis, S.I. A Multi-State Constraint Kalman Filter for Vision-aided Inertial Navigation. In Proceedings of the 2007 IEEE International Conference on Robotics and Automation, Rome, Italy, 10–14 April 2007; Volume 2, p. 6.
6. Qin, T.; Li, P.; Shen, S. Vins-mono: A robust and versatile monocular visual-inertial state estimator. *IEEE Trans. Robot.* **2018**, *34*, 1004–1020. [CrossRef]
7. Mur-Artal, R.; Tardós, J.D. Visual-inertial monocular SLAM with map reuse. *IEEE Robot. Autom. Lett.* **2017**, *2*, 796–803. [CrossRef]
8. Liu, H.; Chen, M.; Zhang, G.; Bao, H.; Bao, Y. Ice-ba: Incremental, consistent and efficient bundle adjustment for visual-inertial slam. In Proceedings of the IEEE Conference on Computer Vision and Pattern Recognition, Salt Lake City, UT, USA, 18–23 June 2018, pp. 1974–1982.
9. Shan, T.; Englot, B.; Meyers, D.; Wang, W.; Ratti, C.; Rus, D. Lio-sam: Tightly-coupled lidar inertial odometry via smoothing and mapping. In Proceedings of the 2020 IEEE/RSJ international conference on intelligent robots and systems (IROS), Las Vegas, NV, USA, 24 October 2020–24 January 2021; pp. 5135–5142.
10. Xu, W.; Cai, Y.; He, D.; Lin, J.; Zhang, F. Fast-lio2: Fast direct lidar-inertial odometry. *IEEE Trans. Robot.* **2022**, *38*, 2053–2073. [CrossRef]

11. Li, K.; Li, M.; Hanebeck, U.D. Towards high-performance solid-state-lidar-inertial odometry and mapping. *IEEE Robot. Autom. Lett.* **2021**, *6*, 5167–5174. [CrossRef]
12. Shan, T.; Englot, B.; Ratti, C.; Rus, D. Lvi-sam: Tightly-coupled lidar-visual-inertial odometry via smoothing and mapping. In Proceedings of the 2021 IEEE international conference on robotics and automation (ICRA), Xi'an, China, 30 May–5 June 2021; pp. 5692–5698.
13. Lin, J.; Zheng, C.; Xu, W.; Zhang, F. R^{2}LIVE: A Robust, Real-Time, LiDAR-Inertial-Visual Tightly-Coupled State Estimator and Mapping. *IEEE Robot. Autom. Lett.* **2021**, *6*, 7469–7476. [CrossRef]
14. Zuo, X.; Geneva, P.; Lee, W.; Liu, Y.; Huang, G. Lic-fusion: Lidar-inertial-camera odometry. In Proceedings of the 2019 IEEE/RSJ International Conference on Intelligent Robots and Systems (IROS), Las Vegas, NV, USA, 24 October 2020–24 January 2021; pp. 5848–5854.
15. Qin, T.; Cao, S.; Pan, J.; Shen, S. A general optimization-based framework for global pose estimation with multiple sensors. *arXiv* **2019**, arXiv:1901.03642.
16. Ahmed, H.; Ullah, I.; Khan, U.; Qureshi, M.B.; Manzoor, S.; Muhammad, N.; Shahid Khan, M.U.; Nawaz, R. Adaptive filtering on GPS-aided MEMS-IMU for optimal estimation of ground vehicle trajectory. *Sensors* **2019**, *19*, 5357. [CrossRef] [PubMed]
17. Nazarahari, M.; Rouhani, H. 40 years of sensor fusion for orientation tracking via magnetic and inertial measurement units: Methods, lessons learned, and future challenges. *Inf. Fusion* **2021**, *68*, 67–84. [CrossRef]
18. Bochkovskiy, A.; Wang, C.Y.; Liao, H.Y.M. Yolov4: Optimal speed and accuracy of object detection. *arXiv* **2020**, arXiv:2004.10934.
19. Garcia-Pulido, J.A.; Pajares, G.; Dormido, S.; de la Cruz, J.M. Recognition of a landing platform for unmanned aerial vehicles by using computer vision-based techniques. *Expert Syst. Appl.* **2017**, *76*, 152–165. [CrossRef]
20. Krishna, N.M.; Reddy, R.Y.; Reddy, M.S.C.; Madhav, K.P.; Sudham, G. Object Detection and Tracking Using Yolo. In Proceedings of the 2021 Third International Conference on Inventive Research in Computing Applications (ICIRCA), Coimbatore, India, 2–4 September 2021; pp. 1–7.
21. Shi, Y.; Qayyum, S.; Memon, S.A.; Khan, U.; Imtiaz, J.; Ullah, I.; Dancey, D.; Nawaz, R. A modified Bayesian framework for multi-sensor target tracking with out-of-sequence-measurements. *Sensors* **2020**, *20*, 3821. [CrossRef] [PubMed]
22. Gong, Z.; Lin, H.; Zhang, D.; Luo, Z.; Zelek, J.; Chen, Y.; Nurunnabi, A.; Wang, C.; Li, J. A Frustum-based probabilistic framework for 3D object detection by fusion of LiDAR and camera data. *ISPRS J. Photogramm. Remote Sens.* **2020**, *159*, 90–100. [CrossRef]
23. Shi, J.; et al. Good features to track. In Proceedings of the IEEE Conference on Computer Vision and Pattern Recognition, Seattle, WA, USA, 21–23 June 1994; pp. 593–600.
24. Lucas, B.D.; Kanade, T. *An Iterative Image Registration Technique with an Application to Stereo Vision*; Morgan Kaufmann Publishers Inc.: San Francisco, CA, USA, 1981; Volume 81.
25. Havlík, J.; Straka, O. Performance evaluation of iterated extended Kalman filter with variable step-length. *J. Phys. Conf. Ser.* **2015**, *659*, 012022. [CrossRef]
26. Ahmadi, M.; Khayatian, A.; Karimaghaee, P. Orientation estimation by error-state extended Kalman filter in quaternion vector space. In Proceedings of the SICE Annual Conference 2007, Takamatsu, Japan, 17–20 September 2007; pp. 60–67.
27. Kaess, M.; Johannsson, H.; Roberts, R.; Ila, V.; Leonard, J.J.; Dellaert, F. iSAM2: Incremental smoothing and mapping using the Bayes tree. *Int. J. Robot. Res.* **2012**, *31*, 216–235. [CrossRef]
28. Lin, J.; Zhang, F. R 3 LIVE: A Robust, Real-time, RGB-colored, LiDAR-Inertial-Visual tightly-coupled state Estimation and mapping package. In Proceedings of the 2022 International Conference on Robotics and Automation (ICRA), Philadelphia, PA, USA, 23–27 May 2022; pp. 10672–10678.
29. Zhu, P.; Wen, L.; Du, D.; Bian, X.; Fan, H.; Hu, Q.; Ling, H. Detection and Tracking Meet Drones Challenge. *IEEE Trans. Pattern Anal. Mach. Intell.* **2021**. [CrossRef] [PubMed]

Article

STV-SC: Segmentation and Temporal Verification Enhanced Scan Context for Place Recognition in Unstructured Environment

Xiaojie Tian [1], Peng Yi [1,2,*], Fu Zhang [3], Jinlong Lei [1,2] and Yiguang Hong [1,2]

[1]   Department of Control Science and Engineering, Tongji University, Shanghai 201804, China
[2]   Shanghai Research Institute for Intelligent Autonomous Systems, Shanghai 201210, China
[3]   Department of Mechanical Engineering, Hong Kong University, Hong Kong 999077, China
*   Correspondence: yipeng@tongji.edu.cn

Abstract: Place recognition is an essential part of simultaneous localization and mapping (SLAM). LiDAR-based place recognition relies almost exclusively on geometric information. However, geometric information may become unreliable when faced with environments dominated by unstructured objects. In this paper, we explore the role of segmentation for extracting key structured information. We propose STV-SC, a novel segmentation and temporal verification enhanced place recognition method for unstructured environments. It contains a range image-based 3D point segmentation algorithm and a three-stage process to detect a loop. The three-stage method consists of a two-stage candidate loop search process and a one-stage segmentation and temporal verification (STV) process. Our STV process utilizes the time-continuous feature of SLAM to determine whether there is an occasional mismatch. We quantitatively demonstrate that the STV process can trigger false detections caused by unstructured objects and effectively extract structured objects to avoid outliers. Comparison with state-of-art algorithms on public datasets shows that STV-SC can run online and achieve improved performance in unstructured environments (Under the same precision, the recall rate is 1.4~16% higher than Scan context). Therefore, our algorithm can effectively avoid the mismatching caused by the original algorithm in unstructured environment and improve the environmental adaptability of mobile agents.

Keywords: place recognition; loop closure; simultaneous localization and mapping (SLAM); unstructured objects; point cloud segmentation; temporal verification

Citation: Tian, X.; Yi, P.; Zhang, F.; Lei, J.; Hong, Y. STV-SC: Segmentation and Temporal Verification Enhanced Scan Context for Place Recognition in Unstructured Environment. *Sensors* **2022**, *22*, 8604. https://doi.org/10.3390/s22228604

Academic Editor: Gregor Klancar

Received: 12 October 2022
Accepted: 3 November 2022
Published: 8 November 2022

Publisher's Note: MDPI stays neutral with regard to jurisdictional claims in published maps and institutional affiliations.

1. Introduction

As the first step towards the realization of autonomous intelligent systems, simultaneous localization and mapping (SLAM) has attracted much interest and made astonishing progress over the past 30 years [1]. Place recognition or loop closure detection gives SLAM the ability to identify previously observed places, which is critical for back-end pose graph optimization to eliminate accumulated errors and construct globally consistent maps [2,3]. Benefiting from the popularity of cameras and the development of computer vision, vision-based place recognition has been widely studied. However, cameras inevitably struggle to cope with illumination variance, poor light conditions, and view-point change [4]. Compared with camera, LiDAR is robust to such perceptual variance and provides stable loop closures. Thus, LiDAR-based recognition has drawn more attention recently. LiDAR-based place recognition is achieved by encoding descriptors directly from geometric information or segmented objects. Then, similarity is assessed by the distance between descriptors, such as multi-view 2D projection (M2DP) [5], bag of words (BOW) [6], scan context (SC) [7], pointnetvlad [8], and overlapTransformer [9]. Descriptors are extracted from local or global geometric information (3D point clouds). Segmatch [10], semantic graph based place recognition [11], semantic scan context (SSC) [12], and RINet [13] leverage the segmented objects to define descriptors.

In this paper, we define relatively large regular objects as structured objects (buildings, ground, trunks, etc.) and others as unstructured objects (vegetation, small moving objects, noise points, etc.). In fact, vegetation is most likely to appear on a large scale and obscure structured information. Thus, we mainly consider unstructured scenes dominated by vegetation. One key issue faced by the above methods is that outliers will occur when two places show similar features due to large scale vegetation. As shown in Figure 1, large scale tree leaves will significantly increase the similarity of different places and reduce the influence of other critical objects in the scene, resulting in similar descriptors between different places. This type of unstructured environment often causes perception aliasing and limits recall rate. Finally, the SLAM system is severely distorted, and the mobile agent cannot perceive the environment correctly. Therefore, designing a place recognition algorithm that is robust in unstructured-dominated environments is of great importance for enhancing the environmental adaptability of autonomous intelligent systems (such as self-driving vehicles and mobile robots) and promoting the development of autonomous driving, field survey, etc.

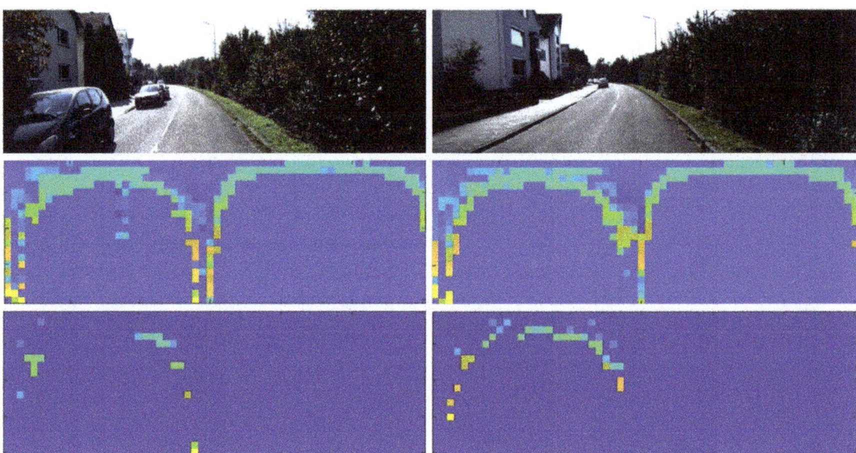

Figure 1. Example of false positive detected by Scan context and triggered by our temporal verification module. Top figures: frame 4058 and 4180 of KITTI sequence 00. The vegetation on the right side makes them difficult to distinguish. Since the ground truth distance between these two frames is 148.64 m, they should not be considered as loop closure. Middle figures: colormap corresponding to scan context before segmentation. Bottom figures: segment scan context of corresponding frame represented by colormap. The left side of colormap indicates the preserved buildings, and the empty right side indicates that the vegetation has been removed. After segmentation, these two frames become distinguishable. If we directly use Scan context, the distance between them is 0.1488, resulting in false positive. Our segment scan context acquires a distance up to 0.327, thus, avoiding outliers.

In [14], segmentation is first proposed to deal with certain conditions, such as forest, and demonstrates potential for removing non-critical information. Inspired by this, here, we intend to enhance scan context with segmentation to make it suitable for unstructured environments. At the same time, considering the time continuity of SLAM and the occasionality of outliers, we use a piecewise thought. Specifically, temporal verification is exploited to candidate loop to decide whether to trigger re-identification module. Thus, reducing the time consumption of the whole system.

In this paper, we present segmentation and temporal verification enhanced scan context (STV-SC). We first design a range image-based segmentation method. Next, we explain why segmented point clouds can differentiate between structured and unstructured objects. Then a three-stage search process is proposed for effective false positives avoidance. The STV process checks temporal consistency to determine whether triggering re-identification module. If triggered, we will segment point clouds and remove unstructured objects of the matching frames. Finally, outliers will be filtered out by the similarity score recomputed by segmented descriptors.

The main contributions of this paper are as follows:

- We propose a range image-based 3D point cloud segmentation method introducing both geometry and intensity constraints for unstructured objects removal.
- An efficient three-stage loop detection algorithm for fast loop candidate search is proposed while leveraging the STV process for perception aliasing rejection.
- Thorough experiments on KITTI dataset [15] show that our method outperforms scan context and other state-of-the-art approaches. The algorithm is also integrated to a SLAM system to verify online place recognition ability.

This paper is structured as follows. Section 2 reviews the related literature of place recognition in both vision and LiDAR manners. Section 3 introduces the 3D point cloud segmentation algorithm proposed, followed by segment scan context and three-stage search algorithm. Then, the experimental test and its discussion are described in Section 4. Finally, a conclusion is made in Section 5.

2. Related Works

Depending on the sensing devices used, place recognition can be grouped into vision-based and LiDAR-based methods. Visual place recognition has been well researched and made significant advancement in the past. Generally, visual approaches represent scene features by extracting multiple descriptors, such as Oriented Fast and Rotated BRIEF (ORB) [16] and Scale-Invariant Feature Transforms (SIFT) [17], to construct a dictionary and then leverage bag of words (BOW) [6] model to measure distance between words that belong to different frames. Recently, a learning-based approach has been used for loop detection [18,19]. NetVlad [18] designed a new generalized VLAD layer and implemented it into CNN to achieve end-to-end place recognition. DOOR-SLAM [20] has verified this method in real world SLAM system. However, image representation usually leads to performance degradation when encountering scenes with light illumination and viewpoint change. To overcome such issues, researchers intended to develop robust visual place recognition methods [21–23] to fit change light and season. In spite of this, these methods can only handle certain scenes.

Unlike a camera, LiDAR is robust to environmental changes stated before, while being rotation-invariant. Now, LiDAR-based recognition is still an advanced and challenging problem for laser SLAM systems. LiDAR methods can be further categorized into local descriptors, global descriptors, and learning-based descriptors. Fast point feature histogram (FPFH) [24], keypoint voting [25], and Combination of Bag of Words and Point Feature [6] are state-of-art approaches based on local hand-crafted descriptors. FPFH [24] is coded by calculating key points and their neighbors' underlying surface properties, such as normal and curvature. Through reordering dataset and caching previously computed values, FPFP can reduce run time and apply to real-time systems. Wang et al. [25] proposed a new 3D regional descriptors based on gestalt features and then certain number of neighbors will be voted by key points to do a similarity score. Bastian et al. [6] used Normal-Aligned Radial Features to build a dictionary for bag of words model and realized robust key points and scene matching.

However, local descriptors rely on the acquisition of key points and the calculation of geometric features around key points, which usually lose a lot of information and lead to false matching. Especially for unstructured outdoor objects (e.g., trees), key points from such objects are unreliable.

In contrast, global descriptors are independent of key points and leverage the global point clouds. Multi-view 2D projection (M2DP) [5] is a novel global descriptor from multi-view 2D mapping of 3D point cloud. This descriptor is designed by the left and right singular vectors of each mapping's density signature. Giseop et al. [7] divided the 3D space into 2D bins and coded each bin by the maximum height of points in this bin. Then, the global descriptor is represented as a two-dimensional matrix called Scan context. The matching of frames is performed by calculating the cosine distance between scan context in column-wise way. Scan context outperforms existing global descriptors and shows remarkable rotation invariance, which allows it to handle reverse loops. Based on scan context, ref. [26] explored the value of intensity. By integrating both geometry and intensity information, they developed intensity scan context and proved that intensity can reflect information of different objects. Meanwhile, they proposed a binary search process, which reduces the computation time significantly.

In recent years, learning based methods have been proposed gradually. Segmatch [10] first segments different objects from original point clouds and then extracts multiple features from each object, such as eigenvalue and shape histograms. Finally, they utilized a learning-based classifier to matching objects of different scenes. Kong et al. [11] leveraged semantic segmentation to build a connected graph by the center of different objects and used CNN network to match scenes by judging the similarity of graphs. Refs. [12,27] proposed semantic scan context, which encodes each bin by semantic information. However, learning-based method is usually computationally expensive for the training process and cannot adapt to various outdoor environments due to the limitation of training data.

Global descriptors show excellent performance, but still cannot handle ambiguous environment caused by unstructured objects and generate outliers. In this paper, inspired by [14], we utilize segmentation to remove unstructured objects of scenes, but remain global information and key structured objects. Then we apply segmented point clouds to scan context and construct segment scan context, which makes different places more distinguishable and effectively prevents perceptual aliasing.

3. Materials and Methods

3.1. System Overview

An overview of the proposed framework is demonstrated in Figure 2. First, the system acquires original 3D point clouds from LiDAR and codes it into scan context. Then, sub-descriptor is designed and put into KD-Tree, which is an indexed tree data structure used for nearest neighbor search in large-scale high-dimensional data spaces. A fast k-Nearest Neighbor (kNN) search is then implemented to find nearest candidates from KD-Tree. Then, by calculating minimum distance between query scan context and candidate scan contexts, we can tell whether there is a candidate loop closure. If it exists, our STV process is conducted. The temporal verification will determine whether to trigger re-identification procedure. Finally, once the temporal verification is met, we consider it to be a true loop. Otherwise, we will segment the original point cloud and then use the segmented scan context to calculate new distance. The re-identification procedure utilizes this distance to judge whether a loop is found. The detailed description of these modules is given below.

3.2. Segmentation

The segmentation module includes two submodules, ground removal and object segmentation. Scan context encodes each bin by taking the maximum height, hence ground points are usually useless and will lead to the increased similarity of different scenes in flat areas. On the other hand, the presence of numerous unstructured objects, such as trees, grass, and other vegetation, will cover the structured information, generating similar descriptors between different places. Meanwhile, it is evident that noises generally do not persist in a certain position over time. Thus, they generally appear scattered and form small-scale objects. Here, we use object segmentation to remove unstructured information in the environment and retain key structured information to prevent mismatches.

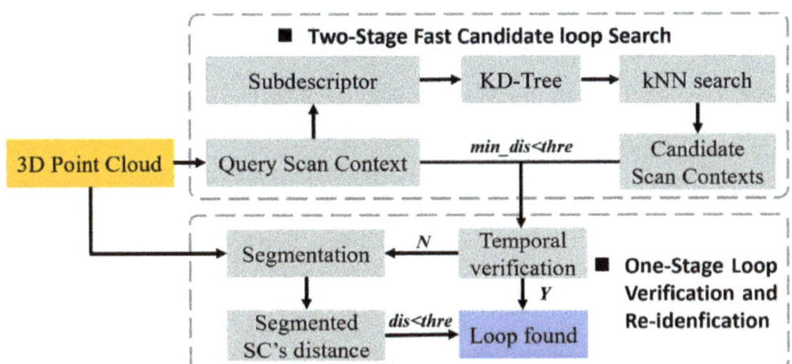

Figure 2. The pipeline of the proposed STV-SC framework. Grey dotted box above: the two-stage fast candidate loop closure search process, including a k-Nearest Neighbor (kNN) search process and a similarity scoring process. Grey dotted box below: our STV process.

We denote each frame of point cloud from the LiDAR as $\mathcal{P} = \{p_1, p_2, \ldots, p_n\}$. For fast cluster-based segmentation, the 3D point cloud is projected into a $M_r \times M_c$ 2D range image R for point cloud ordering, where

$$M_r = \frac{360°}{Res_h} \quad , \quad M_c = N_{scans}. \tag{1}$$

Res_h is the horizontal resolution and N_{scans} is the line number of LiDAR. Each value of the range image is represented by the Euclidean distance from the sensor to the corresponding point cloud in 3D space. Then, we use a column-wise approach for ground point evaluation on the range image like [28], while leveraging intensity for validation.

After removing the ground, we perform a range image-based object segmentation to classify point clouds into distinct clusters, which is based on [29] but with some improvements according to the characteristics of LiDAR. Specifically, we integrate geometry and intensity constraints for clustering. Previous study [30] showed that different objects exhibit different reflected intensity. Since intensity can be obtained directly from LiDAR, it can serve as an additional layer of validation for clustering. We can judge whether two points p_a and p_b belong to object O_k by the following mathematical expression. Meanwhile, we set (a_1, a_2) and (b_1, b_2) as their coordinates in the range image, respectively:

$$p_a, \quad p_b \in O_k$$

$$s.t. \quad ||a_1 - b_1|| = 1 \quad or \quad ||a_2 - b_2|| = 1$$

$$\theta > \epsilon_g$$

$$||I(p_a) - I(p_b)|| < \epsilon_i$$

$$\theta = \arctan \frac{d_2 \sin \gamma}{d_1 - d_2 \cos \gamma}$$

$$I(p) = \kappa(\psi(p), d). \tag{2}$$

In (2), as shown in Figure 3, d stands for the range value from LiDAR to 3D point cloud. θ is the angle between the line spawned by p_a, p_b and the longer one of OA and OB. ϵ_g and ϵ_i are predefined thresholds. Additionally, $\psi(p)$ denotes the intensity of point p and κ is an intensity calibration function using distance, which can be obtained by practice.

Noticed that as the first-layer judgment, geometry constraint plays a major role. As the second-layer of validation, intensity prevents objects of different types from being clustered together, i.e., under-segmentation.

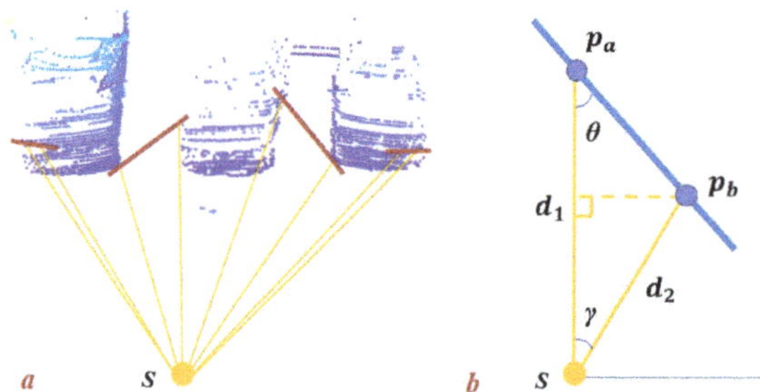

Figure 3. Interpretation of geometry constraint for segmentation. (**a**): three parking cars and laser beams from sensor S. The red line represents the line spawned by two adjacent points. (**b**): geometric abstraction of (**a**). p_a and p_b represent two adjacent points.

Moreover, due to the fixed angle between laser beams, points distributed near the sensor are relatively dense, while points far away are sparse. If a fixed geometric threshold is used, we cannot balance the distant and near points. Specifically, if a large threshold is used, the distant points will be over-segmented, and if a small threshold is used, the nearby points will be under-segmented. Thus, in the near area, using a large ϵ_g can prevent different objects from being grouped together, while using a small ϵ_g in the far area can avoid the same object being segmented into multiple objects.

To achieve more accurate segmentation at different distances, we design a dynamic adjustment strategy. Threshold will be dynamically adjusted as

$$\epsilon_g = \epsilon_g^i - \frac{R(x,y)}{p}q,\tag{3}$$

where p denotes step size and q is the decay factor. ϵ_g^i stands for the initial value of ϵ_g.

Finally, a breadth-first search based on constraints in (2) is conducted on range image for object clustering. The idea of our segmentation comes from the fact that unstructured objects (mainly vegetation) are filled with gaps, such as leaves. When the laser beams pass through the gaps, the range difference will become large, which will cause large scale vegetation to be separated into small clusters. In the meantime, noise is also a small object. Therefore, we can distinguish structured and unstructured objects by the size of the clusters. In this paper, we treat clusters with more than 30 points or occupying over 5 laser beams as structured objects. As shown in Figure 4, noises, ground, and vegetation are removed, while structured parts, such as buildings and parking cars, are preserved.

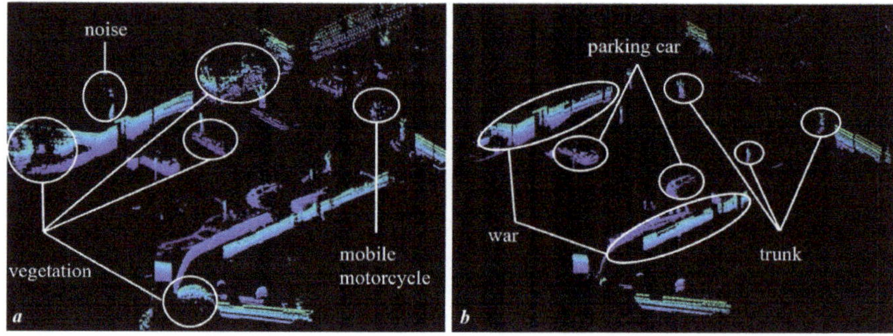

Figure 4. Visualization of the segmentation process. (**a**): original point clouds of one frame. Vegetation, small moving object, and noise are present. (**b**): segmented point clouds, which shows that unstructured vegetation, noise, etc., are removed.

3.3. Segment Scan Context

Scan context [7] encodes the scene with the maximum height and then represents it by a 2D image. Figure 5a is the top view of original point clouds. Taking the LiDAR as the center, N_r rings are equidistantly divided in the radial direction. In the azimuth direction, N_s sectors are divided by equal angles. The area where rings and sectors intersect are called bins. For each bin, a unique representation of the maximum height of point clouds within it is used. Therefore, we can project the 3D point clouds into a 2D matrix of $N_r \times N_s$, called scan context. Let L_{max} represents the maximum sensing range of the LiDAR, then the gaps of rings and sectors are $\dfrac{L_{max}}{N_r}$ and $\dfrac{2\pi}{N_s}$, respectively. By adjusting them, we can set the resolution of scan context.

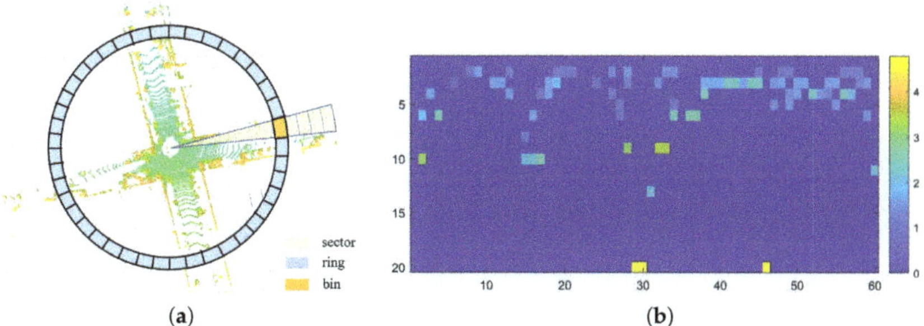

(a) (b)

Figure 5. Description of scan context. (**a**): top view of a LiDAR scan, which is separated into bins by rings and sectors. (**b**): colormap of our segment scan context.

However, since scan context uses the maximum height as the unique encoding, it usually results in perceptual aliasing when facing large scale unstructured objects. Like trees on both sides of road, they usually have the same height. Therefore, when encountering scenes dominated by unstructured objects, we merely maintain key structured information obtained via point cloud segmentation. Denote point clouds of a segmented LiDAR scan as $\mathcal{P}^{seg}$, segment scan context D is expressed by

$$D = (d_{ij}) \in \mathbb{R}^{N_r \times N_s} \quad , \quad d_{ij} = \phi(\mathcal{P}^{seg}_{ij}). \tag{4}$$

$\mathcal{P}^{seg}_{ij}$ are points in a bin with ring index i and sector index j and ϕ denotes the function to obtain the maximum height of all point clouds in this bin. Particularly, if there is no point in the bin, its value is set to zero. Visualization of our segment scan context is in Figure 5b. After segmentation, descriptors exhibit discrete blocks representing different structured objects.

3.4. Three-Stage Search Algorithm

After projecting original point clouds into scan context, the matching process is dedicated to calculating the minimum distance between the descriptor D_t obtained at time t and the $\mathcal{D} = \{D_1, D_2, \ldots, D_{t-1}\}$ stored previously. Then, the distance determines whether there is a loop closure. In order to achieve fast search and effectively prevent mismatches, we design a three-stage search and verification algorithm.

Stage 1: Fast k-Nearest Neighbor search. Obviously, searching in the database directly using scan context will generate numerous decimal operations, which will slow down the search speed. Here, we perform fast search by extracting sub-descriptors. First, scan context is binarized as follows. Let B denotes the matrix after binarization:

$$B(x,y) = \begin{cases} 0, & \text{if } D(x,y) = 0, \\ 1, & \text{otherwise.} \end{cases} \tag{5}$$

Then, for each row r of B, we count the number of non-empty bins by calculating L_0 norm:

$$v(r_i) = \|r_i\|_0. \tag{6}$$

Finally, we construct a one-dimensional sub-descriptor $H = (v(r_1), v(r_2), \ldots, v(r_n))$ that fulfills rotation invariance. By putting H into KD-Tree, we can achieve fast kNN search and provide k candidates for the next stage.

Stage 2: Similarity score with column shift. This step will directly use the corresponding scan context to find the nearest frame from the candidates obtained in stage 1. Let D^q denotes the scan context of query scan. D^c denotes one candidate scan context. A column-wise accumulation of cosine distances is used to measure the distance between D^q and D^c. The distance is:

$$\varphi(D^q, D^c) = \frac{1}{N_s} \sum_{i=1}^{N_s} \left(\frac{c_i^q \cdot c_i^c}{\|c_i^q\| \cdot \|c_i^c\|} \right), \tag{7}$$

where c_i^q and c_i^c are the i-th column of D^q and D^c, respectively. In practice, mobile agents may revisit one place from different view-points. To achieve rotation invariance, we conduct a column shift process as

$$\varphi_{min}(D^q, D^c) = \min_{j \in [1, N_s]} \varphi(D^q, D_j^c), \tag{8}$$

where D_j^c means shift D^c by j columns and φ_{min} represents the final smallest value. If φ_{min} is lower than the predefined threshold ϵ_l, then we obtain a candidate D^c for next stage.

Stage 3: Temporal verification and re-identification (STV process). To effectively prevent the generation of false positives, we design a temporal verification module for this candidate loop. Since the detection process of SLAM is continuous in time, the nodes near a true loop also have high similarity. Furthermore, true loops usually exist continuously, while outliers are sporadic. Therefore, we adopt a piecewise idea to verify candidate loop pair:

$$\mathcal{T}(D_m, D_n) = \frac{1}{N_t} \sum_{k=1}^{N_t} \varphi_{min}(D_{m-k}, D_{n-k}), \tag{9}$$

where N_t means the quantity of frames involved for temporal verification. If $\mathcal{T}$ less than a threshold ϵ_t, we treat it as a true loop. Otherwise, we regard this frame as ambiguous environment and the re-identification module with our segment scan context will be triggered. Specifically, we segment original point clouds and calculate distance between segment scan context of candidate loop pair. Since we have obtained the shift value in the previous stage, we can directly use the result in Equation (8) to calculate the new distance:

$$\varphi_{seg}(D^{segq}, D^{segc}) = \varphi(D^{segq}, D_{j*}^{segc}), \tag{10}$$

where j^* represents the shift value when $\varphi_{min}(D^q, D^c)$ reaches. Finally, if φ_{seg} still less than a threshold ϵ_s, we group it into inliers; otherwise, we discard it.

Algorithm 1 depicts our search process in detail, where num_diff represents the minimum interval between two frames that can become a loop closure. min_dis means minimum distance.

Algorithm 1 Tree-stage search process

Require: Original point cloud $\mathcal{P}$ of current frame at time t.
Require: Scan context D^q of current frame at time t.
Require: Sub-descriptors of the previous frames in KD-Tree.
Require: Previous scan contexts $\mathcal{D}$ stored before t

1: $k \leftarrow 50$, $q \leftarrow$ index of current frame.
2: $num\_diff \leftarrow 50$, $min\_dis \leftarrow 100{,}000$.
3: Build the sub-descriptor H of the current frame (Equations (5) and (6)) and insert it into KD-Tree.
4: **if** $q > k$ **then**
5:     Find k nearest candidates in KD-Tree (kNN search).
6:     **for** $i = 1$ to k **do**
7:         $ii \leftarrow$ frame index of ith candidate.
8:         **if** $ii - q > num\_diff$ **then**
9:             Calculate the distance φ between frame q and ii (Equations (7) and (8)).
10:             **if** $\varphi < min\_dis$ **then**
11:                 $min\_dis \leftarrow \varphi$, $D^c \leftarrow D^{ii}$.
12:             **end if**
13:         **end if**
14:     **end for**
15:     **if** $min\_dis < \epsilon_l$ **then**
16:         Temporal verification of D^q and D^c (Equation (9)).
17:         **if** $\tau < \epsilon_t$ **then**
18:             Loop found!
19:         **else**
20:             Segment $\mathcal{P}$ to get $\mathcal{P}^{seg}$ (Equation (2)).
21:             Construct segment scan context D^{seg} (Equation (4)).
22:             Calculate the distance φ_{seg} between D^{segq} and D^{segc} (Equation (7)).
23:             **if** $\varphi_{seg} < \epsilon_s$ **then**
24:                 Loop found!
25:             **end if**
26:         **end if**
27:     **end if**
28: **end if**

---

4. Experimental Results and Discussion

In this section, we conduct a series of experiments to verify the effectiveness of our STV process for unstructured scenes. Moreover, the discussion regarding each experiment is also presented. The performance of our algorithm is compared with other state-of-art global descriptors. All experiments are performed on a computer equipped with an Intel Core (TM) i5-10210U CPU. To compare with Scan context [7] and test online capability, our algorithm is implemented both in MATLAB and C++.

4.1. Experimental Setup

We select four sequences (00, 05, 06, and 08) from the KITTI dataset [15], all of which contain a large number of typical scenes dominated by unstructured objects (mainly vegetation). As shown in Figure 6, these outdoor scenes provide sufficient experimental resources for our algorithm.

Figure 6. Typical scenes from KITTI sequences. (**a**) sequence 00; (**b**) sequence 05; (**c**) sequence 06; and (**d**) sequence 08. These scenes are dominated by unstructured objects, which can easily cause mismatches.

In order to show higher accuracy and exhibit the application value of the algorithm, our parameter settings are similar to scan context-50 [7]. This means that in the first stage we will select 50 nearest neighbors, while ensuring real-time performance. If the ground truth Euclidean distance of matched pair is less than 4m, we consider it to be an inlier. Since ϵ_l and ϵ_t have the same physical meaning, we make them equal in the experiment. Other parameter values used are listed in Table 1.

Table 1. Parameter List.

Parameter	Value
Maximum radius (L_{max})	80
Number of rings (M_r)	20
Number of sectors (M_s)	60
Segmentation threshold (ϵ_g)	60
Segmentation threshold (ϵ_i)	0.5
Re-identification threshold (ϵ_s)	0.2–0.3
Frames of temporal verification (N_t)	2

4.2. Statistical Analysis

To illustrate that our STV process can increase the distinguishability in scenes with large scale unstructured objects and effectively avoid the occasional mismatches brought by such scenes. We perform a statistical analysis.

The 4000~4400th frames of KITTI sequence 00 contain a lot of places dominated by vegetation. Many of these frames are highly susceptible to mismatches, which are discovered through our temporal verification module.

We first carry out analysis on the structured and unstructured objects of the selected 400 frames to demonstrate that the segmentation module described in Section 3.2 can indeed separate unstructured objects from structured objects. Figure 7 presents our statistical results. We can find that the clustering number of structured objects after segmentation is much less than that of an unstructured one. The mean values in Figure 7a,b demonstrate a difference of about 30 times. We represent the size of a cluster by the number of points included. Figure 7c,d show that the former tend to be larger clusters, while the latter are small in size due to gaps in vegetation or noises. Generally, structured clusters are more than 10 times larger than unstructured clusters. Therefore, we naturally think of using the size of the cluster to remove vegetation, etc. In subsequent experiments, we will retain clusters with more than 30 points or occupying over 5 laser beams as structured objects.

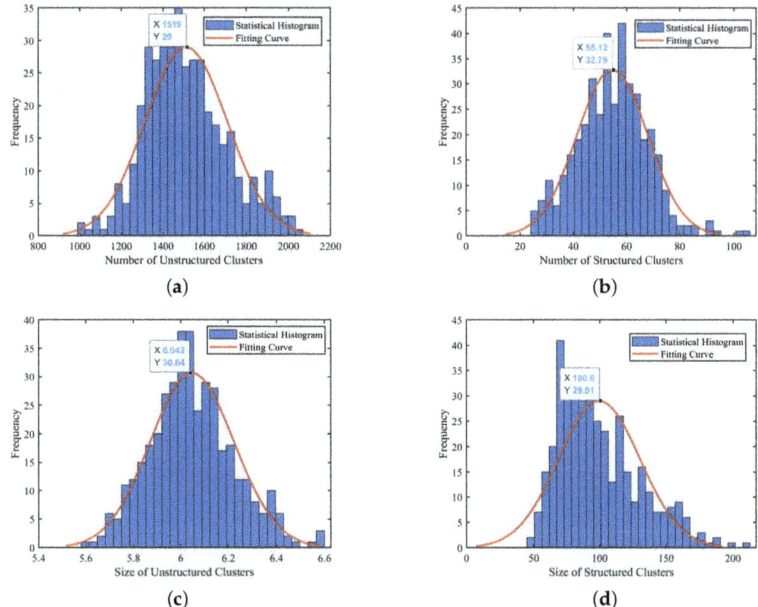

Figure 7. Comparison of segmentation results between structured and unstructured objects. (**a**,**b**): Distribution of the number of unstructured and structured clusters in each frame, respectively. (**c**,**d**): Distribution of the average size of unstructured and structured clusters in each frame, respectively.

Second, we compare the similarity scores of these 400 pairs of false positives before and after segmentation. As shown in Figure 8, the scores between different places are significantly increased after removing the unstructured objects, as vegetation always has a high degree of similarity. It means improved distinguishability between false loop closures. This allows our algorithm to directly discard mismatches when encountering such places.

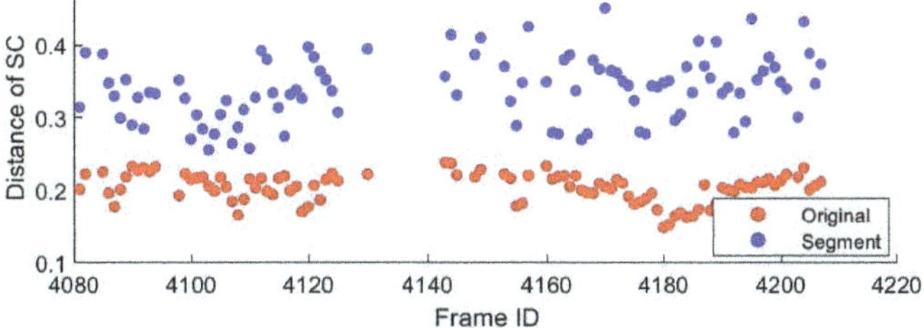

Figure 8. Comparison of false loop pairs' similarity scores before and after segmentation.

4.3. Dynamic Threshold Evaluation

In our segmentation algorithm, as the first step judgment, the geometric threshold plays a more critical role in accurate segmentation. According to the characteristics of laser beams, we design a dynamic adjustment strategy of ϵ_g, as shown in Equation (3), which can prevent under-segmentation of near objects and over-segmentation of far objects compared with the fixed geometric threshold.

Here, we use the control variable method to test the influence of the dynamic threshold on the experimental results, so as to provide a parameter reference for next experiment. Specifically, we compare the precision and recall rates of fixed and dynamic thresholds with different initial values of ϵ_g. Experiments are performed on KITTI sequences 00 and 08, which can provide more convincing references due to their large number of complex and typical unstructured scenes. From the results in Table 2, we can see that under the same initial value, the dynamic threshold tends to achieve higher recall and precision rates than the fixed one. Moreover, we can conclude that the initial value of ϵ_g is best set between 50 and 60.

Table 2. Precision and recall rates of different ϵ_g, p and q.

Parameter			Sequence 00		Sequence 08	
ϵ_g^i (°)	p	q	Precision	Recall	Precision	Recall
50	-	-	0.875	0.653	0.998	0.916
55	-	-	0.880	0.707	0.998	0.916
55	20	0.5	0.881	0.711	0.998	0.916
60	-	-	0.894	0.714	0.946	0.912
60	10	1	0.915	0.714	0.998	0.916
65	-	-	0.720	0.714	0.934	0.919
65	10	1	0.809	0.714	0.948	0.918

Therefore, in the following experiments, we set parameters of dynamic threshold as $\epsilon_g^i = 60$, $p = 10$ and $q = 1$.

4.4. Precision Recall Evaluation

We leverage precision-recall curves to comprehensively evaluate the performance of our STV-SC method in environments where large scale unstructured objects exist. The performance of our place recognition algorithm is compared with Scan context [7] and M2DP [5], since both are state-of-art global descriptors and neither specifically considers unstructured scenes. In particular, our algorithm is enhanced from Scan context, so the performance comparison with Scan context in unstructured environments is quite important.

As shown in Figure 9, the experiments are conducted on sequences 00, 05, 06, and 08. Since sequence 08 only has reverse loop, it can verify that our algorithm maintains the rotation invariance of Scan context.

Our proposed algorithm outperforms other approaches in all sequences. This is because in the suburban where the roads are surrounded by trees, the geometric information for place recognition is limited. For example, the frames we mentioned in Section 4.2, Scan context will cause mismatches due to the existence of vegetation. However, our method can mitigate the impact of vegetation and avoid many mismatches caused by unstructured objects. That is, under the same recall rate, STV-SC can obtain higher precision rate. As for sequence 08, M2DP performs poorly due to its inability to achieve rotation invariance. However, our algorithm achieves improved performance while maintaining rotation invariance. The residual outliers come from jungles with few or no structured objects or scenes where the structured parts are still very similar so that the geometric information can no longer meet the requirements of place recognition.

In the application, we pay more attention to the recall rate under high precision. Table 3 shows the recall of sequences 00, 05, and 06 at 100% precision. Since sequence 08 is more challenging, we take the recall rate when the precision is 90%. It is obvious that our method outperforms other approaches which do not consider unstructured objects. Compared with the original Scan context, the recall rate of our STV-SC algorithm on different sequences is increased by 1.4% to 16%. In particular, in sequence 08, an environment with a lot of vegetation. Other algorithms often have poor performance, while our algorithm improves the recall rate by more than 15%.

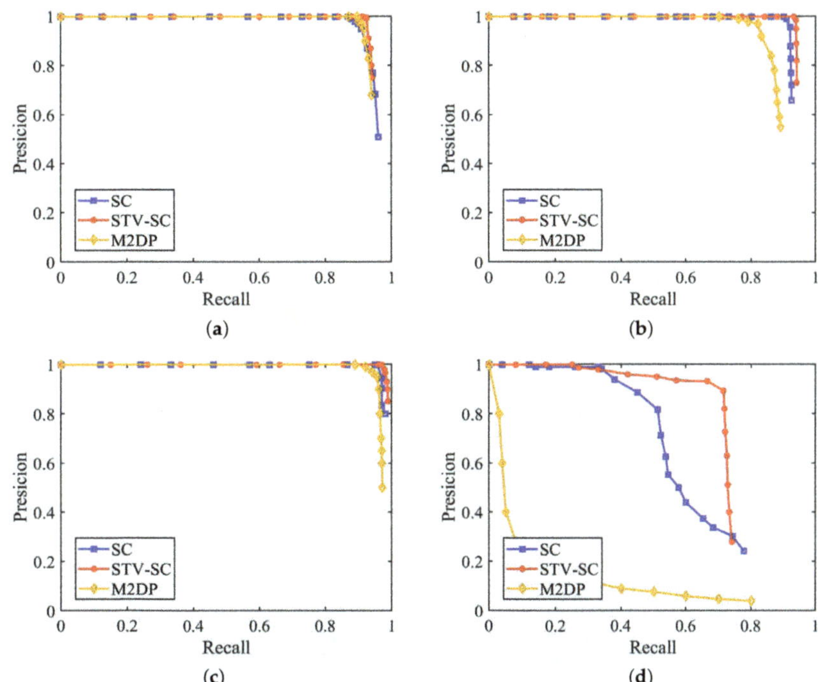

Figure 9. Precision-recall curves on KITTI dataset. (**a**) sequence 00; (**b**) sequence 05; (**c**) sequence 06; and (**d**) sequence 08. The performance of the algorithms is measured by the area enclosed by the curves and the coordinate axes.

Table 3. Recall at 100% precision on KITTI 00, 05, and 06; Recall at 90% precision on KITTI 08.

	Sequence 00		Sequence 05		Sequence 06		Sequence 08	
Methods	Precision	Recall	Precision	Recall	Precision	Recall	Precision	Recall
Scan Context	1.000	0.870	1.000	0.900	1.000	0.956	0.900	0.550
STV-SC	1.000	0.912	1.000	0.931	1.000	0.970	0.900	0.714
M2DP	1.000	0.896	1.000	0.761	1.000	0.890	0.900	0.020

4.5. Time-Consumption Analysis

Compared to Scan context, our method adds segmentation and temporal verification (STV) process. Since the re-identification module does not require search and shift actions, the main time consumption of STV is concentrated in the segmentation module. In the meantime, subject to temporal verification, re-identification process is not always triggered, but only used when encountering ambiguous environment. As the main time-consuming module, segmentation uses a range image-based breadth-first search, whose time consumption is fairly small.

Under the same conditions as Scan context-50 [7], we record the place recognition time consumption (cost time of STV-SC) for more than 100 triggered frames in Figure 10. Even at the peak, the time consumption is less than 0.4 s. The average time consumption of these 120 frames is 0.316 s (the original scan context is 0.201 s under 0.2 m^3 point cloud downsampling), which is within a reasonable range (2–5 HZ on Matlab).

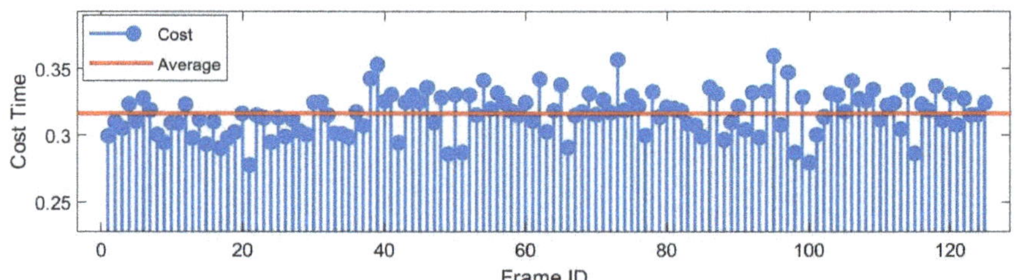

Figure 10. Time-consumption result of 120 triggered frames on KITTI 00. In the case of triggering re-identification, the average time consumption of the whole system is 0.316 s.

4.6. Online Loop-Closure Performance

Now, we show the online performance of our STV-SC algorithm. Our algorithm is integrated into the well-known LiDAR odometry framework LOAM [31]. Specifically, our method is used as the loop closure detection module of LOAM, then the detected loop is added to the pose graph as an edge. GTSAM [32] is applied for back-end graph optimization. Finally, a drift-free trajectory is obtained. The experiments run on Robot Operating System (ROS Melodic) and perform on KITTI 00.

The white dots in Figure 11 represent examples of detected loop closures. As shown in the estimated trajectory, our method can effectively detect loop closures and eliminate drift errors in real time, even in unstructured-dominated environments.

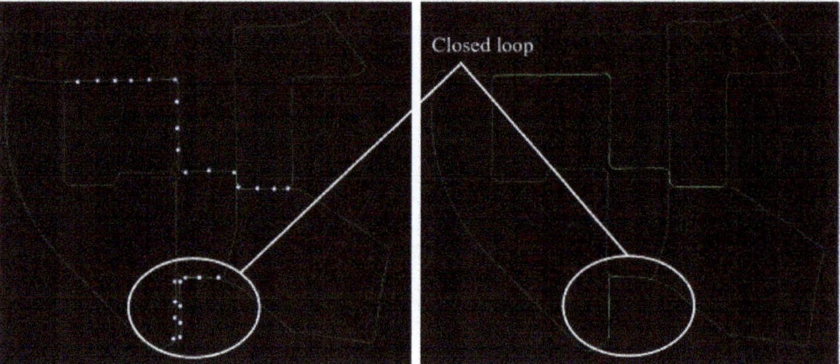

Figure 11. Online loop-closure performance of STV-SC on KITTI 00. Left figure shows the trajectory without loop closure detection and pose graph optimization. The trajectory in the white circle exhibits noticeable drifts. Right figure shows the trajectory after pose graph optimization.

5. Conclusions

In this paper, we have proposed STV-SC, a new Scan context-based place recognition method that integrates segmentation and temporal verification process, which gives the original algorithm the ability to handle unstructured environments and enhances the stability of mobile agents in special and complex environments. By summarizing the characteristics of unstructured objects, we design a novel segmentation method to distinguish unstructured and structured objects according to the size of clusters. In addition, for more accurate segmentation we adopt a geometric threshold that varies with range value. In the matching part, we design a three-stage algorithm. Based on the temporal continuity of SLAM system, if temporal verification is not satisfied, the re-identification module will be triggered. Thus, effectively avoiding mismatches caused by unstructured objects. Comprehensive experiments on the KITTI dataset demonstrate that our segmentation method can effectively distinguish different types of objects. STV-SC achieves higher recall and precision rates than Scan context and other state-of-art global descriptors in vegetation-dominated environments. Specifically, it is considered that under the same precision, the recall rate can be improved by 1.4~16% by our algorithm in different datasets. Meanwhile, the average time consumption of STV-SC is 0.316 s which is within a reasonable bound and proves that the our algorithm can be run in the SLAM system online.

Author Contributions: Conceptualization, X.T. and P.Y.; methodology, X.T.; software, X.T.; validation, X.T., P.Y., F.Z. and Y.H.; formal analysis, X.T.; investigation, X.T.; resources, X.T. and J.L.; data curation, X.T.; writing—original draft preparation, X.T.; writing—review and editing, X.T., P.Y., F.Z., J.L. and Y.H.; visualization, X.T.; supervision, P.Y., J.L. and Y.H.; project administration, P.Y. and J.L.; funding acquisition, P.Y. and J.L. All authors have read and agreed to the published version of the manuscript.

Funding: This research was sponsored by Shanghai Sailing Program under Grant Nos. 20YF1453000 and 20YF1452800, the National Natural Science Foundation of China under Grant Nos. 62003239 and 62003240, and the Fundamental Research Funds for the Central Universities under Grant Nos. 22120200047 and 22120200048. This work has been supported in part by Shanghai Municipal Science and Technology, China Major Project under grant 2021SHZDZX0100 and by Shanghai Municipal Commission of Science and Technology, China Project under grant 19511132101.

Institutional Review Board Statement: Not applicable.

Informed Consent Statement: Not applicable.

Data Availability Statement: Not applicable.

Conflicts of Interest: The authors declare no conflict of interest.

Abbreviations

The following abbreviations are used in this manuscript:

SLAM	Simultaneous Localization and Mapping
STV	Segmentation and Temporal Verification
SC	Scan Context
LOAM	Lidar Odometry and Mapping

References

1. Cadena, C.; Carlone, L.; Carrillo, H.; Latif, Y.; Scaramuzza, D.; Neira, J.; Reid, I.; Leonard, J.J. Past, Present, and Future of Simultaneous Localization and Mapping: Toward the Robust-Perception Age. *IEEE Trans. Robot.* **2016**, *32*, 1309–1332. [CrossRef]
2. Saeedi, S.; Trentini, M.; Seto, M.; Li, H. Multiple-robot simultaneous localization and mapping: A review. *J. Field Robot.* **2016**, *33*, 3–46. [CrossRef]
3. Cattaneo, D.; Vaghi, M.; Valada, A. Lcdnet: Deep loop closure detection and point cloud registration for lidar slam. *IEEE Trans. Robot.* **2022**, *38*, 2074–2093. [CrossRef]
4. Taketomi, T.; Uchiyama, H.; Ikeda, S. Visual SLAM algorithms: A survey from 2010 to 2016. *IPSJ Trans. Comput. Vis. Appl.* **2017**, *9*, 16. [CrossRef]
5. He, L.; Wang, X.; Zhang, H. M2DP: A novel 3D point cloud descriptor and its application in loop closure detection. In Proceedings of the 2016 IEEE/RSJ International Conference on Intelligent Robots and Systems (IROS), Vancouver, BC, Canada, 24–28 September 2016; pp. 231–237.
6. Steder, B.; Ruhnke, M.; Grzonka, S.; Burgard, W. Place recognition in 3D scans using a combination of bag of words and point feature based relative pose estimation. In Proceedings of the 2011 IEEE/RSJ International Conference on Intelligent Robots and Systems, San Francisco, CA, USA, 25–30 September 2011; pp. 1249–1255.
7. Kim, G.; Kim, A. Scan context: Egocentric spatial descriptor for place recognition within 3d point cloud map. In Proceedings of the 2018 IEEE/RSJ International Conference on Intelligent Robots and Systems (IROS), Madrid, Spain, 1–5 October 2018; pp. 4802–4809.
8. Uy, M.A.; Lee, G.H. Pointnetvlad: Deep point cloud based retrieval for large-scale place recognition. In Proceedings of the IEEE Conference on Computer Vision and Pattern Recognition, Salt Lake City, UT, USA, 18–22 June 2018; pp. 4470–4479.
9. Ma, J.; Zhang, J.; Xu, J.; Ai, R.; Gu, W.; Chen, X. OverlapTransformer: An Efficient and Yaw-Angle-Invariant Transformer Network for LiDAR-Based Place Recognition. *IEEE Robot. Autom. Lett.* **2022**, *7*, 6958–6965. [CrossRef]
10. Dubé, R.; Dugas, D.; Stumm, E.; Nieto, J.; Siegwart, R.; Cadena, C. Segmatch: Segment based place recognition in 3d point clouds. In Proceedings of the 2017 IEEE International Conference on Robotics and Automation (ICRA), Singapore, 29 May–3 June 2017; pp. 5266–5272.

11. Kong, X.; Yang, X.; Zhai, G.; Zhao, X.; Zeng, X.; Wang, M.; Liu, Y.; Li, W.; Wen, F. Semantic graph based place recognition for 3d point clouds. In Proceedings of the 2020 IEEE/RSJ International Conference on Intelligent Robots and Systems (IROS), Las Vegas, NV, USA, 24 October 2020–24 January 2021; pp. 8216–8223.

12. Li, L.; Kong, X.; Zhao, X.; Huang, T.; Li, W.; Wen, F.; Zhang, H.; Liu, Y. SSC: Semantic scan context for large-scale place recognition. In Proceedings of the 2021 IEEE/RSJ International Conference on Intelligent Robots and Systems (IROS), Prague, Czech Republic, 27 September–1 October 2021; pp. 2092–2099.

13. Li, L.; Kong, X.; Zhao, X.; Huang, T.; Li, W.; Wen, F.; Zhang, H.; Liu, Y. RINet: Efficient 3D Lidar-Based Place Recognition Using Rotation Invariant Neural Network. *IEEE Robot. Autom. Lett.* **2022**, *7*, 4321–4328. [CrossRef]

14. Shan, T.; Englot, B. Lego-loam: Lightweight and ground-optimized lidar odometry and mapping on variable terrain. In Proceedings of the 2018 IEEE/RSJ International Conference on Intelligent Robots and Systems (IROS), Madrid, Spain, 1–5 October 2018; pp. 4758–4765.

15. Geiger, A.; Lenz, P.; Urtasun, R. Are we ready for autonomous driving? the kitti vision benchmark suite. In Proceedings of the 2012 IEEE Conference on Computer Vision and Pattern Recognition, Providence, RI, USA, 16–21 June 2012; pp. 3354–3361.

16. Rublee, E.; Rabaud, V.; Konolige, K.; Bradski, G. ORB: An efficient alternative to SIFT or SURF. In Proceedings of the 2011 International Conference on Computer Vision, Barcelona, Spain, 6–13 November 2011; pp. 2564–2571.

17. Lowe, D.G. Object recognition from local scale-invariant features. In Proceedings of the Seventh IEEE International Conference on Computer Vision, Kerkyra, Corfu, Greece, 20–25 September 1999; pp. 1150–1157.

18. Arandjelovic, R.; Gronat, P.; Torii, A.; Pajdla, T.; Sivic, J. NetVLAD: CNN architecture for weakly supervised place recognition. In Proceedings of the IEEE Conference on Computer Vision and Pattern Recognition, Las Vegas, NV, USA, 27–30 June 2016; pp. 5297–5307.

19. Kendall, A.; Grimes, M.; Cipolla, R. Posenet: A convolutional network for real-time 6-dof camera relocalization. In Proceedings of the IEEE International Conference on Computer Vision, Santiago, Chile, 7–13 December 2015; pp. 2938–2946.

20. Lajoie, P.-Y.; Ramtoula, B.; Chang, Y.; Carlone, L.; Beltrame, G. DOOR-SLAM: Distributed, online, and outlier resilient SLAM for robotic teams. *IEEE Robot. Autom. Lett.* **2020**, *5*, 1656–1663. [CrossRef]

21. Anoosheh, A.; Sattler, T.; Timofte, R.; Pollefeys, M.; Van Gool, L. Night-to-day image translation for retrieval-based localization. In Proceedings of the 2019 International Conference on Robotics and Automation (ICRA), Montreal, QC, Canada, 20–24 May 2019; pp. 5958–5964.

22. Milford, M.J.; Wyeth, G.F. SeqSLAM: Visual route-based navigation for sunny summer days and stormy winter nights. In Proceedings of the 2012 IEEE International Conference on Robotics and Automation, St. Paul, MN, USA, 14–18 May 2012; pp. 1643–1649.

23. Arshad, S.; Kim, G.-W. An Appearance and Viewpoint Invariant Visual Place Recognition for Seasonal Changes. In Proceedings of the 2020 20th International Conference on Control, Automation and Systems (ICCAS), Busan, Korea, 13–16 October 2020; pp. 1206–1211.

24. Rusu, R.B.; Blodow, N.; Beetz, M. Fast point feature histograms (FPFH) for 3D registration. In Proceedings of the 2009 IEEE International Conference on Robotics and Automation, Kobe, Japan, 12–17 May 2009; pp. 3212–3217.

25. Bosse, M.; Zlot, R. Place recognition using keypoint voting in large 3D lidar datasets. In Proceedings of the 2013 IEEE International Conference on Robotics and Automation, Karlsruhe, Germany, 6–10 May 2013; pp. 2677–2684.

26. Wang, H.; Wang, C.; Xie, L. Intensity scan context: Coding intensity and geometry relations for loop closure detection. In Proceedings of the 2020 IEEE International Conference on Robotics and Automation (ICRA), Paris, France, 31 May–31 August 2020; pp. 2095–2101.

27. Li, Y.; Su, P.; Cao, M.; Chen, H.; Jiang, X.; Liu, Y. Semantic Scan Context: Global Semantic Descriptor for LiDAR-based Place Recognition. In Proceedings of the 2021 IEEE International Conference on Real-time Computing and Robotics (RCAR), Xining, China, 15–19 July 2021; pp. 251–256.

28. Himmelsbach, M.; Hundelshausen, F.V.; Wuensche, H.-J. Fast segmentation of 3D point clouds for ground vehicles. In Proceedings of the 2010 IEEE Intelligent Vehicles Symposium, La Jolla, CA, USA, 21–24 June 2010; pp. 560–565.

29. Bogoslavskyi, I.; Stachniss, C. Fast range image-based segmentation of sparse 3D laser scans for online operation. In Proceedings of the 2016 IEEE/RSJ International Conference on Intelligent Robots and Systems (IROS), Daejeon, Korea, 9–14 October 2016; pp. 163–169.

30. Kashani, A.G.; Olsen, M.J.; Parrish, C.E.; Wilson, N. A review of LiDAR radiometric processing: From ad hoc intensity correction to rigorous radiometric calibration. *Sensors* **2015**, *15*, 28099–28128. [CrossRef] [PubMed]

31. Zhang, J.; Singh, S. LOAM: Lidar odometry and mapping in real-time. In Proceedings of the Robotics: Science and Systems, University of California, Berkeley, CA, USA, 12–16 July 2014; pp. 1–9.

32. Kaess, M.; Johannsson, H.; Roberts, R.; Ila, V.; Leonard, J.J.; Dellaert, F. iSAM2: Incremental smoothing and mapping using the Bayes tree. *Int. J. Robot. Res.* **2012**, *31*, 216–235. [CrossRef]

sensors

Article

A Method of Calibration for the Distortion of LiDAR Integrating IMU and Odometer

Qiuxuan Wu [1,*], Qinyuan Meng [1], Yangyang Tian [2], Zhongrong Zhou [1], Cenfeng Luo [1], Wandeng Mao [2], Pingliang Zeng [1], Botao Zhang [1] and Yanbin Luo [1]

[1] School of Automation, Hangzhou Dianzi University, Hangzhou 310018, China
[2] Electric Power Research Institute of State Grid Henan Electric Power Company, Zhengzhou 450018, China
* Correspondence: wuqx@hdu.edu.com

Abstract: To improve the motion distortion caused by LiDAR data at low and medium frame rates when moving, this paper proposes an improved algorithm for scanning matching of estimated velocity that combines an IMU and odometer. First, the information of the IMU and the odometer is fused, and the pose of the LiDAR is obtained using the linear interpolation method. The ICP method is used to scan and match the LiDAR data. The data fused by the IMU and the odometer provide the optimal initial value for the ICP. The estimated speed of the LiDAR is introduced as the termination condition of the ICP method iteration to realize the compensation of the LiDAR data. The experimental comparative analysis shows that the algorithm is better than the ICP algorithm and the VICP algorithm in matching accuracy.

Keywords: motion distortion; IMU; sensor fusion; odometer; ICP

Citation: Wu, Q.; Meng, Q.; Tian, Y.; Zhou, Z.; Luo, C.; Mao, W.; Zeng, P.; Zhang, B.; Luo, Y. A Method of Calibration for the Distortion of LiDAR Integrating IMU and Odometer. *Sensors* **2022**, *22*, 6716. https://doi.org/10.3390/s22176716

Academic Editors: Yong Liu and Xingxing Zuo

Received: 25 July 2022
Accepted: 2 September 2022
Published: 5 September 2022

Publisher's Note: MDPI stays neutral with regard to jurisdictional claims in published maps and institutional affiliations.

1. Introduction

Exact pose estimation is the key technology for mapping, location, and navigation in the field of the mobile robot [1], which can provide the message of the robot's position and gesture in real time. Sensors used to obtain robot pose estimates include LiDAR, cameras, wheel encoders, IMUs, etc. According to the different sensors the robot is equipped with, SLAM technology is divided into visual SLAM and laser SLAM. Although the sensor used in visual SLAM has low cost and rich image information, it has a great impact on the normal operation of the camera under weak- or no-light conditions. What is more, because the image information is too rich, the algorithm requires high processor performance. Now, the mainstream mobile robots are still dominated by laser sensors [2], such as the unmanned delivery vehicle of JD and the "prime" unmanned delivery vehicle of Amazon.

A 2D LiDAR estimates the pose of the sensor by scan-matching two adjacent frames of laser data [3]. However, only relying on 2D laser SLAM to estimate the pose of the robot has many limitations. The frequency of the system output estimated pose is low, and the running time becomes longer, which will generate a large cumulative error and eventually affects the positioning and map construction of the robot. A cartographer algorithm [4] is developed using a SICK radar, and the frame rate can reach more than 100 Hz. The motion distortion can be ignored, so there is no distortion correction algorithm module. However, the frame rate of most LiDAR is around 10 Hz. Without distortion correcting, there will be distortion error appearing in LiDAR data, which is hard to eliminate through loopback detection and back-end optimization, etc. The research on this issue has great practical significance. Many domestic and foreign works have been conducted on removing motion distortion and false match of LiDAR data in recent years. Yoon et al. [5] proposed an unsupervised parameter learning in the Gaussian variational inference setting, which combines classical trajectory estimation of mobile robots and deep learning on rich sensor data to learn a complete estimator via the deep network. However, it requires a large amount of calculation, the captured laser data cannot complete feature extraction or matching when the environment is not clearly structured, and the real-time and robustness are poor. Therefore, it is only suitable for small indoor scenes with clear structure instead of open large outdoor scenes. Hyeong et al. [6] proposed an ICP (Iterative Closest Points, iterative closest point) outlier rejection scheme to compare the laser data of the scanned environment and select matching points and reject the algorithm that does not match parts. The ICP algorithm needs to be provided with an initial value, and the matching accuracy of the ICP algorithm directly depends on whether the initial value is accurate. However, in the process of acquiring the surrounding environment, the laser is often accompanied by the motion of the robot. Especially when the laser frame rate is small, the captured laser data will produce motion distortion, and there will be a large error with the real environment over time. Xue et al. [7] proposed a simultaneous fusion of IMU, wheel encoder, and LiDAR to estimate the own motion of a moving vehicle. However, this method does not propose a countermeasure for discontinuous laser scanning. Bazet and Cherfaoui [8] proposed a method for correcting errors caused by time stamp errors during sensor data acquisition, but this scheme assumes that the scanning angle of the laser is fixed and the quadratic interpolation assumption is too simplistic, which cannot meet the complex outdoor environment. Hong et al. [9] proposed a new approach to enhancing ICP algorithms by updating speed, which estimates the speed of the LiDAR through ICP iterations, and uses the estimated speed to compensate for scan distortion due to motion. Although it considers the motion of the robot into consideration, its assumption of uniform motion is too ideal; for low-frame rate LiDAR, the assumption of uniform motion does not hold.

Aiming at the above problems, this paper proposes an improved algorithm for estimated speed scan matching that integrates an IMU and odometer. This algorithm is called Iao_ICP (ICP that integrates IMU and Odometer) in this paper. The main contributions of this paper are as follows: (1) The algorithm uses the linear interpolation method to obtain the pose of LiDAR, which solves the alignment problem of the discontinuous laser scan data. (2) The data fused by the IMU and the odometer provides a better initial value for the ICP, and the estimated speed of the LiDAR is introduced as the iterative value of the ICP method to realize the termination condition of LiDAR data compensation.

The rest of the paper is organized as follows: Firstly, the causes of motion distortion in the traditional ICP algorithm are analyzed. Secondly, the incremental information of the wheel odometer and the angular velocity information of the IMU are integrated into the pose estimation. Finally, through data sets and physical experiments, the effectiveness of the proposed algorithm in removing motion distortion and improving the accuracy of map construction is demonstrated.

2. Causes of LiDAR Motion Distortion

The mechanical LiDAR is driven by an internal motor to rotate the radar ranging core $360°$ clockwise to obtain the surrounding environment data. Each frame of laser data is encapsulated by the data information obtained by a certain number of discrete laser beams, and the laser data of each frame is not obtained instantaneously. The data distortion of LiDAR is related to the motion state of the robot which carries LiDAR. When laser scanning is accompanied by the motion of the robot, the laser data of each angle is not obtained instantaneously. When the scanning frequency of the LiDAR is relatively low, the motion distortion of the laser frame caused by the motion of the robot cannot be ignored [10].

The current domestic LiDAR rotation frequency is about 5–10 Hz. When the robot carrying the LiDAR is stationary, the measurement data of the LiDAR has no error, but in the SLAM system, the robot is often in a state of motion. Take the environment shown in Figure 1 as an example. It can be seen that the distance data of each laser beam are collected in different poses, as shown in the pose of points A and B. Suppose the robot is moving at a constant speed, the solid curved arrow indicates that the LiDAR rotation direction is clockwise, and the solid long straight arrow indicates that the LiDAR moves from point A to point B along the X direction. Then, in the case of no motion distortion correction during this period, the LiDAR data will have a motion distortion error of Δx.

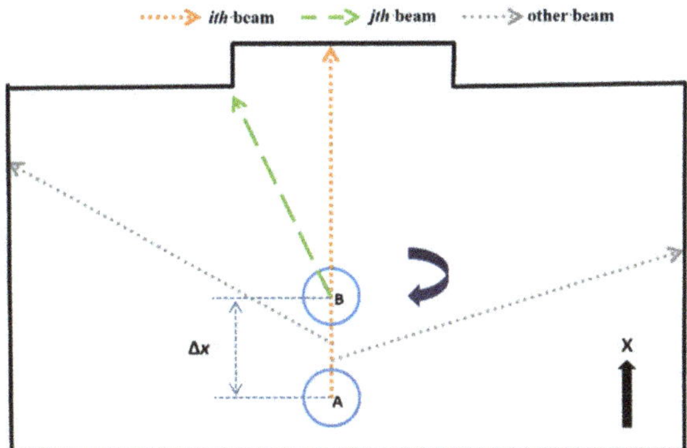

Figure 1. The acquisition process of one frame of LiDAR data.

As described above, when the robot obtained a frame of LiDAR data, the laser is obtained at point A, and the laser is obtained at point B. However, when general LiDAR drives package data, it is assumed that all laser beams of a frame of LiDAR data are obtained in the same pose and instantaneously, that is, all laser beams are obtained from point A data. Its pose actually produces motion changes, and each laser point is generated on a different reference pose, which eventually causes the environmental distortion of the laser collection. As Figure 2 shows, the left picture is the actual environment, while the dotted line in the picture on the right is the true value, and the solid line is the LiDAR data with motion distortion.

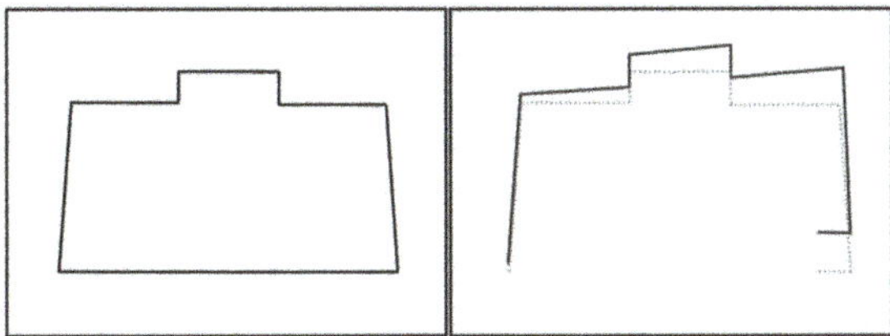

Figure 2. LiDAR motion distortion.

3. Principle of ICP Algorithm

The ICP algorithm [11] was first developed by Beals and McKay in 1992. The ICP algorithm is essentially an optimal registration method based on the least-squares method. ICP first matches each point of the target laser data with the closest point of the reference laser data and finds the rotation matrix R and translation matrix p, which are used to convert the two. Afterward, the laser matching is iteratively optimized by repeatedly generating pairwise closest points until the convergence accuracy requirements for correct registration are met. The ICP algorithm first needs to determine an initial pose, and the selected initial value will have an important impact on the final registration result. The algorithm may fall into a local optimum instead of a global minimum if the initial value is not chosen properly.

Given $X = \{x_1, x_2, \cdots, x_{N_x}\}$ as a frame of laser data, $P = \left\{p_1, p_2, \cdots, p_{N_p}\right\}$ as the laser data of adjacent frames, and $T = \{T_1, T_2, \cdots, T_i\}$ as the transformation matrix of laser data of adjacent frames, x_i and p_i indicate the coordinates of the laser spot, N_x and N_p indicate the number of laser dots, and i indicates the frame number of laser data. This paper defined a minimizing objective Function (1) to transform P through the coordinates, and cover the maximum to X [11].

$$E(R, p) = \frac{1}{N_p} \sum_{i=1}^{N_p} \| x_i - Rp_i - t \|^2 \tag{1}$$

The resulting transformation matrix T can be described as (2):

$$T = \begin{bmatrix} R & p \\ 0 & 1 \end{bmatrix}$$

(2)

The processing steps of the given objective function are shown as follows:
Step1: Solving the mean value of LiDAR data X and P:

$$U_x = \frac{1}{N_x} \sum_{i=1}^{N_x} X_i, U_p = \frac{1}{N_p} \sum_{i=1}^{N_p} P_i;$$

Step2: Remove the translation of LiDAR data X and P to distributed laser data around the mean value:

$$x_i' = x_i - U_x, p_i' = p_i - U_p$$

Step3: Define matrix, and make SVD decomposition of it, where H is the matrix to be decomposed by SVD, U and V are the two non-singular matrices decomposed, and σ_1, σ_2, and σ_3 are the three singular values decomposed, respectively:

$$H = \sum_{i=1}^{Np} x_i' p_i'^T = U \begin{bmatrix} \sigma_1 & 0 & 0 \\ 0 & \sigma_2 & 0 \\ 0 & 0 & \sigma_3 \end{bmatrix} V^T$$

Step4: Calculate the solution of the objective function:

$$R = UV^T, \ p = u_x - Ru_p$$

Since the ICP algorithm uses the closest point as the corresponding point, the initial result may be different from the real environment. However, the results converge to the base environment by repeating this process. The LiDAR scan data for frame i, namely, X, are shown in Figure 3a. The LiDAR scan data for frame $i + 1$, namely, P, are shown in Figure 3b. The first step of ICP iteration is shown in Figure 3c. The closest point between X and P is found as Figure 3d shows. The first matching estimated transformation and updated P by $p_i' = T_1 p_i$, which is shown in Figure 3e. The X and P matched after many iterations, as Figure 3f shows. Final pose estimation is solved through the transformation of $T = T_n T_{n-1} \cdots T_2 T_1 (i = 1, \cdots, n)$, namely:

$$x_i = T_n T_{n-1} \cdots T_2 T_1 p_i = T p_i$$

(3)

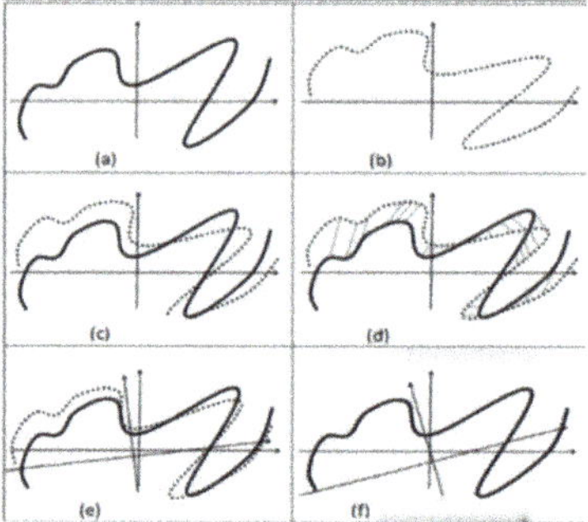

Figure 3. The principle of ICP algorithm. (**a**) Frame i (**b**) Frame $i + 1$ (**c**) Start matching (**d**) Find adjacent (**e**) First match (**f**) After multiple iterations of matching.

4. Estimation Speed Scan Matching Algorithm Based on IMU and Odometer

A wheeled odometer and IMU are introduced to compensate for motion distortion of laser data caused by robot moves. Direct measurement of displacement and angle information through a wheel odometer or direct measurement of angular velocity and linear acceleration through an IMU [12], then integrate them, respectively, to obtain the displacement and angle information. In the ideal conditions, the wheel odometer or IMU has a high-precision local pose estimation ability because of the high pose update frequency (higher than 200 Hz) of the above sensors, which can accurately reflect the motion of the robot in real time [13]. What is more, these two types of sensors are completely decoupled from the robot state estimation, which can prevent the introduction of errors. However, on the one hand, during the actual movement of the robot, the wheels will slip and the accumulated error will occur, which leads to a certain deviation in the obtained odometer angle data when only the encoder is used, and the error increases with the running time and the stroke increases. On the other hand, the linear acceleration accuracy of the IMU is poor, though it has high angular velocity measurement accuracy, and the local accuracy of the quadratic integral is still very poor, which leads to a certain deviation of obtained displacement data. Therefore, this paper proposes the Iao_ICP algorithm, and the algorithm framework is shown in Figure 4. First, the information of the IMU and the odometer is fused, and the pose of the LiDAR is obtained using the linear interpolation method to remove most of the motion distortion. Then, scan matching of LiDAR data is conducted using the ICP method. Data fused by the IMU and odometer provide a better initial value for ICP, and estimated speed is introduced as a termination condition for iteration of the ICP method [14]. The matching result is used as the correct value, and the error value of the odometer is obtained. The error value is evenly distributed to each point, and the position of the laser point is corrected again, so as to further determine the pose of the laser point.

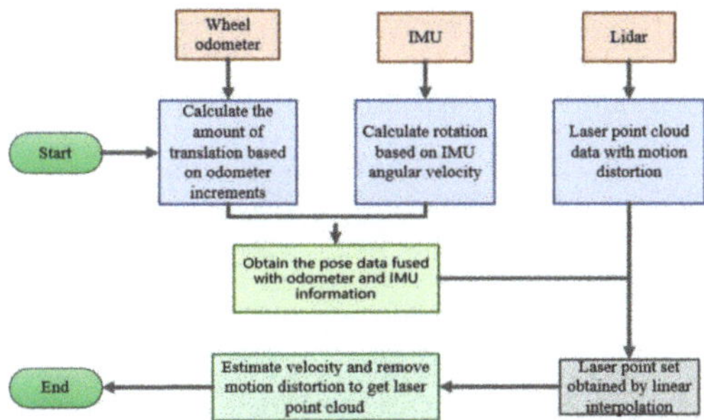

Figure 4. The architecture diagram of the Iao_ICP algorithm.

4.1. Pose Estimation with Fusion of IMU and Odometer

The chassis control system of the mobile robot reads the IMU data and the odometer data. Each time the IMU data are read, the odometer data can also be obtained without considering the problem of time synchronization. That means the IMU pose queue and odometer pose queue maintain strict alignment, which can directly fuse both to generate a new pose queue. However, the update frequency of low-cost LiDAR is generally only 5–10 Hz, which leads to the new pose queue after fusion cannot maintain strict alignment with the pose queue of laser frames. Although there is no way to obtain the pose of the laser frame directly from the fused pose queue since the pose queues of the two are not strictly aligned, the pose of the laser frame can be obtained by linearly interpolating the fused pose queue. Below are the detailed steps to obtain the estimated pose based on the linear interpolation method by fusing the IMU and odometer data:

Step1: As the start time of the current laser frame, the end time of the current laser frame, and the time interval between two laser beams have been known. Odometer data and IMU data are stored in a queue in the same chronological order, and the team leader is the earliest. There are oldest odometer and IMU data timestamps, and latest odometer and

IMU data timestamps. First, solve the new queue generated by fusing odometry and IMU data within the above timestamps. The fusion expressions are shown below:

$$\begin{cases} Odom\_Imu\_List[i].x = OdomList[i].x \\ Odom\_Imu\_List[i].y = OdomList[i].y \\ Odom\_Imu\_List[i].\theta = ImuList[i].\theta \end{cases} \tag{4}$$

In the formula, $Odom\_Imu\_List[i]$ is the fused pose data at the t_i moment, $OdomList[i]$ is the odometer pose data at the t_i moment, $ImuList[i]$ is the IMU pose data at the t_i moment, and x, y, and θ are the X-axis data, Y-axis data, and angle data in the pose data, respectively.

Step2: Solve the emission pose corresponding to each laser in the current frame of laser data, namely, to solve the robotic pose at the time of $\{t_s, t_s + \Delta t, \cdots t_s + i\Delta t \cdots t_e\}$. It is reasonable to assume that the robot moves at a uniform speed during the data update of the fusion of two adjacent frames due to the high update frequency of odometer data and IMU data. Linear interpolation can be used on this assumption, as shown in Figure 5.

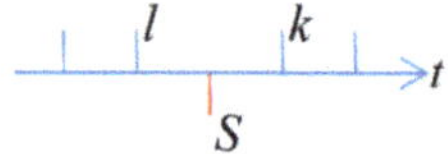

Figure 5. Linear interpolation of laser pose.

Suppose there are corresponding fused pose queues at the time of l, k for laser data, but not at the time of s, and the value of s is greater than l, and less than k. Then, solve the pose of robot p_s, p_m, p_e corresponding to the three moments t_s, t_m, t_e ($t_s < t_m < t_e$). The pose of the first laser beam can be calculated with the Formula (5). In the same way, the emission pose of the last laser beam and the laser beam at the middle time can be obtained.

$$\begin{cases} p_l = Odom\_Imu\_List[i] \\ p_k = Odom\_Imu\_List[k] \\ p_s = p_l + \frac{p_k - p_l}{k - l}(s - l) \end{cases} \tag{5}$$

Step3: Following the method in the Step2, p_m and p_e can be solved. Further assumed, the robot performs uniform acceleration motion during a frame of laser data. Thus, the pose of the robot is a quadratic function of time, as Figure 6 shows. Thus, using the known robot pose p_s, p_m, p_e as the independent variable, a quadratic curve function $P(t) = At^2 + Bt + C(t_s < t < t_e)$ can be obtained by interpolation, and A, B, C are the coefficients of the quadratic function. Next, the value of every time $\{t_s, t_s + \Delta t, \cdots t_s + i\Delta t \cdots t_e\}$ can be substituted into a curve, and the pose of each laser point data in global coordinate system $\{p_{t_s}, p_{t_s+\Delta t}, \cdots p_{t_s+i\Delta t} \cdots p_{t_e}\}$ can be obtained.

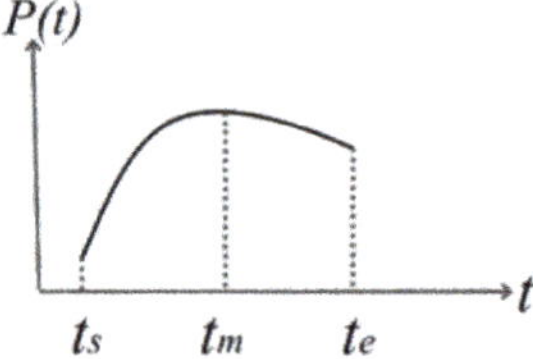

Figure 6. Pose function graph.

Step4: The relative pose (array form) of the laser point in the global coordinate system is converted into a pose change matrix. Then, convert the coordinate information in the radar coordinate system xi to the coordinates in the global coordinate system, as Formula (6) shows.

$$x_i' = V2T(p_i)x_i \tag{6}$$

In the above Formula (6), function $V2T\ (p_i)$ is a whole, indicating that the relative pose in the form of an array p_i is converted into a pose transformation matrix in the form of a matrix. By the coordinate information in radar system x_i left multiplication corresponding matrix p_i, the coordinate of the radar coordinate system x_i can be translated into the coordinate in the global coordinate system x_i', because p_i is the pose in the global coordinate system.

Step5: According to the coordinates of the scanning point corresponding to each laser beam in the global coordinate system x_i', the laser data of the laser scan point in the LiDAR coordinate system can be solved with Formula (7).

$$\begin{cases} x_i' = (p_x, p_y) \\ range = \sqrt{p_x \cdot p_x + p_y \cdot p_y} \\ angle = \text{atan2}(p_y, p_x) \end{cases} \tag{7}$$

For the first equation above, p_x, p_y are the coordinates of the ith frame of laser data in the LiDAR coordinate system on the X- and Y-axis, respectively.

For the second equation above, the coordinates p_x and p_y on the x and y axes of the laser x_i frame in the laser coordinate system are known. The distance point x_i from the origin of the laser coordinate system can be found according to the "Pythagorean Theorem".

For the third equation above, p_x and p_y have been found, and the angle between point x_i and X-axis can be solved according to inverse trigonometric functions. The specific implementation process of the algorithm is shown in Algorithm 1.

Algorithm 1: A Pose Estimation Algorithm Based on IMU and Odometer

---

Input: Odometer pose queue $OdomList[i]$, IMU pose queue, and laser pose queue x_i
Output: laser pose queue X_n

1: **for** $i = 1{:}n$ **do**

2: $Odom\_Imu\_List[i].x = OdomList[i].x;$
$Odom\_Imu\_List[i].y = OdomList[i].y;$
$Odom\_Imu\_List[i].\theta = ImuList[i].\theta;$ //fuse the data of odometer and IMU pose queue, then put into $Odom\_Imu\_List[i]$

3: **end for**

4: $p_s = LinerInterp(Odom\_Imu\_List[t_s]);$
$p_m = LinerInterp(Odom\_Imu\_List[t_m]);$
$p_e = LinerInterp(Odom\_Imu\_List[t_e]);$ //Perform linear interpolation on the fusion pose of the start, end and intermediate moments, $LinerInterp()$ is function used to make linear interpolation

5: $P(t) = P(t) = At^2 + Bt + C;$ //Substitute p_s, p_m, p_e into above formula in order, and the coefficients of quadratic curve functions A, B, C can be solved.

6: **for** $i = 1{:}n$ **do**

7: $p_i = Ai^2 + Bi + C;$ //solve the pose of each laser point in global coordinate system p_i

8: $x'_i = V2T(p_i)x_i = (p_x, p_y);$ //obtain the pose of each laser point in the global coordinate system x'_i

9: $X_n = (range, angle) = \left(\sqrt{p_x * p_x + p_y * p_y}, \text{atan2}(p_y, p_x) \right);$ //compose a new laser point set X_n

10: **end for**

4.2. Estimated Velocity and Laser Data Pose Compensation

To remove the motion distortion of the laser point cloud data, the speed of the robot needs to be estimated. Since the scanning period of LiDAR is about 0.1 s, it can be assumed that the speed of the robot is constant during this scanning period, and V_i is used to indicate the velocity in the LiDAR coordinate system at t_i time. Firstly, estimate the velocity V_i from the relative motion transformation between two adjacent frames of laser data X_i and X_{i-1}, supposing that n indicates the number of laser points of laser data X_i. The time interval between two adjacent frames of laser points is Δt. x_0, x_1, $\cdots$, x_n is the laser point of laser X_i, $t_{x_j} - t_{x_{j-1}} = \Delta t_s$ $(j = 0, 1, \cdots, n-1)$.

Therefore, the estimated velocity V_i is:

$$V_i = \frac{T2V\left(T_{i-1}^{-1}T_i\right)}{\Delta t} \approx \frac{1}{\Delta t}\lg T_{i-1}^{-1}T_i \tag{8}$$

In Formula (8), $T_{i-1}^{-1}T_i$ is a whole, indicating the pose difference of the robot from $i-1$ time to i time in the radar coordinate system, and $T2V\left(T_{i-1}^{-1}T_i\right)$ indicates a way to convert the pose difference from matrix form to array form.

The pose of frame i and laser point j is:

$$T(t_i + j\Delta t_s) = T_i \cdot V2T(V_i \cdot j\Delta t_s) = T_i e^{j\Delta t_s}V_i \tag{9}$$

In Formula (9), $j\Delta t_s$ is the duration of laser point cloud data in frame i from laser point 0th to laser point j.

V_i is the origin velocity of laser point data in frame i.

$V_i \cdot j\Delta t_s$ is the pose difference of frame i laser data cloud from laser point 0th to laser point k.

$V2T(V_i \cdot j\Delta t_s)$ is the conversion of relative pose difference from array form to matrix form.

$T_i \cdot V2T(V_i \cdot j\Delta t_s)$ is to obtain the pose of laser point j in frame i by using frame i of laser point cloud data right-multiplied by the pose difference from the initial pose of the 0th laser point.

Substitute the above Formula (9) into Formula (3), the laser point cloud data collection X_i is converted into $\overline{X}^*$, and $\overline{X}^*$ is the laser point cloud data collection after speed compensation.

$$\overline{X}^* = \left\{ e^{j\Delta t_s V_i} p_j \mid j = 0, 1, \cdots, n \right\} \tag{10}$$

For some types of LiDAR, it takes 100 ms to perform a scan with a scan angle of 360°, which takes the estimation of robot motion later than the actual movement. To prevent this kind of delay, a backward compensation scheme can be used. Take the time corresponding to the last laser point as the reference time, the corresponding time of each laser point can be converted. With the above conditions, Formula (9) can be revised into:

$$T[t_i - (n-j)\Delta t_s] = T_i e^{(n-j)\Delta t_s(-V_i)} \tag{11}$$

Formula (10) can be revised into:

$$\overline{X} = \left\{ e^{(n-j)\Delta_s(-V_i)} x_j \mid j = 0, 1, \cdots, n \right\} \tag{12}$$

The specific implementation process of the algorithm is shown in Algorithm 2.

Algorithm 2: Estimating velocity and removing motion distortion from laser point cloud data combined with ICP

Input: the queue of laser pose X_n
Output: motion transformation matrix of adjacent laser frames T

1: $V = V_i$ //speed initialization

2: **do**

3: $T_{\Delta ts} = e^{\Delta ts(-V_i)}$ //the motion transformation matrix T is estimated by the speed of the two adjacent frames of laser light

4: **for** $j = 1 : n$ **do** //traverse all laser points in the current laser frame

5: $T_{j\Delta ts} = T_{(j-1)\Delta ts}T_{\Delta ts}$ //calculate the motion transformation matrix of each laser point

6: $\bar{x}_{i_j} = T_{j\Delta ts}x_{i_j}$ //Motion transformation for each laser point

7: end **for**

8: $T = ICP\left(\bar{X}^{-1}, \bar{X}_i, T\right)$ //iterative matching via ICP

9: $V = V_i$ //renew the value of velocity

10: $V_i = 1/\Delta lg\ T$ //do the next round of speed estimation

11: ***While*** $||V - V_i|| > e$ //when the speed error value is greater than the threshold e, execute the loop

---

5. Positioning Accuracy Evaluation of Laser Odometry after Motion Distortion Calibration

This experiment utilizes the sequences b0_2014_07_11_10_58_16 (denoted as ①), b0_2014_07_11_11_00_49 (denoted as ②), and b0_2014_07_21_12_42_53 (denoted as ③) in the Cartographer public dataset. The laser odometry accuracy of the Iao_ICP algorithm and the original Cartographer algorithm is quantitatively evaluated by executing this. Figure 7 shows the mapping effect of the Iao_ICP algorithm on the sequence. The processor of the test equipment is Intel (R) Core (TM) i5−5200 CPU 2.20 GHz and it has 8 GB RAM.

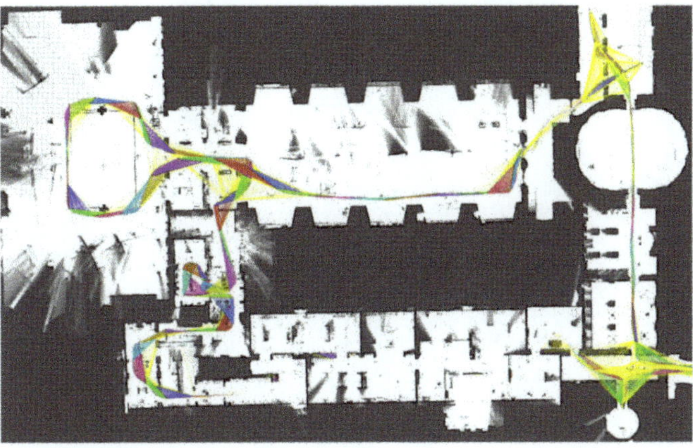

Figure 7. Mapping based on the b0_2014_07_11_11_00_49 sequence.

The analysis is performed by comparing the data calculated by the Iao_ICP algorithm with the Cartographer data set. Table 1 lists the absolute trajectory errors calculated by these two algorithms [15]. In addition, the Iao_ICP algorithm was used to calculate the relative trajectory error and compared with the relative trajectory error of the original Cartographer algorithm.

Table 1. Comparison results of absolute trajectory error between Iao_ICP algorithm and Cartographer algorithm.

Sequence	Algorithm	RMSE (m)	Average (m)	Maximum (m)	Minimum (m)
①	Iao_ICP	0.0179	0.0103	0.01356	0.0011
	Cartographer	0.023	0.0147	0.01428	0.0024
②	Iao_ICP	0.0166	0.0091	0.1510	0.0008
	Cartographer	0.0197	0.0096	0.1663	0.0011
③	Iao_ICP	0.0158	0.0089	0.1233	0.0003
	Cartographer	0.0193	0.0092	0.1349	0.0012

Using sequence ① for testing, the comparison of the relative trajectory error results obtained is shown in Figure 8.

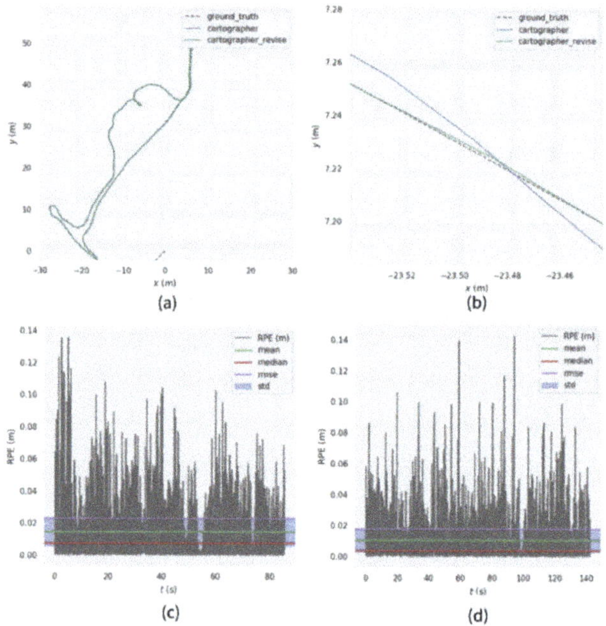

Figure 8. Comparison of relative trajectory errors of sequence ①. (**a**) Cartographer, improvement scheme, and real trajectory comparison (**b**) Local trajectory map (**c**) Absolute trajectory error of Cartographer (**d**) Absolute trajectory error of the improved scheme.

The obtained comparison of relative trajectory error results is shown in Figure 9 by using a sequence for testing.

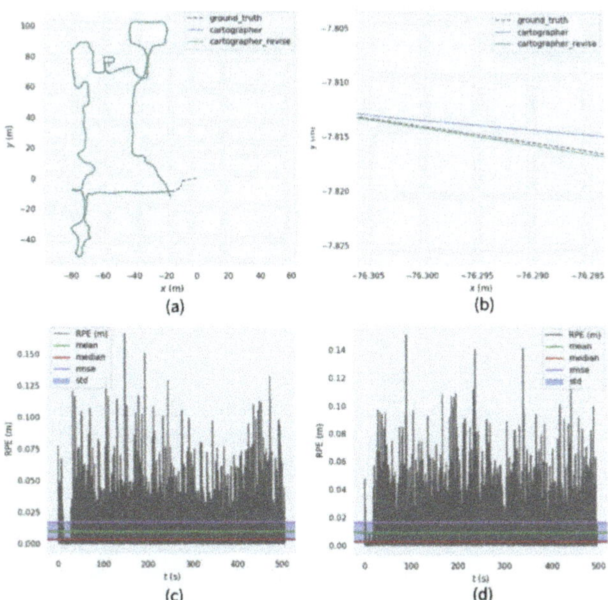

Figure 9. Comparison of relative trajectory errors of sequence ②. (**a**) Cartographer, improvement scheme, and real trajectory comparison (**b**) Local trajectory map (**c**) Absolute trajectory error of Cartographer (**d**) Absolute trajectory error of the improved scheme.

The obtained comparison of relative trajectory error results is shown in Figure 10 by using a sequence for testing.

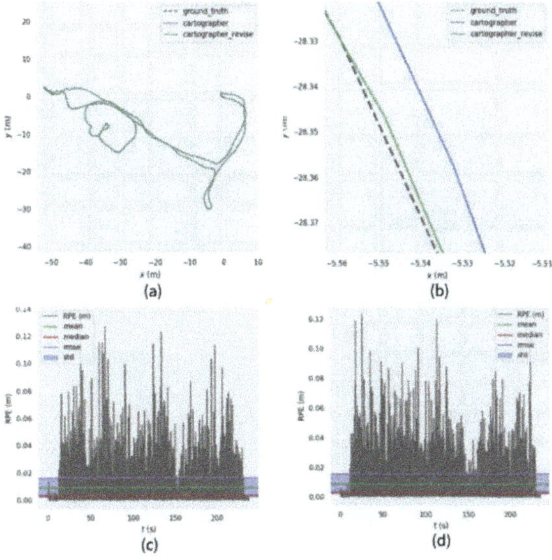

Figure 10. Comparison of relative trajectory errors of sequence ③. (**a**) Cartographer, improvement scheme, and real trajectory comparison (**b**) Local trajectory map (**c**) Absolute trajectory error of Cartographer (**d**) Absolute trajectory error of the improved scheme.

From the sequence ① test results, it can be seen that, from the RMSE index, the root-mean-square error of the Iao_ICP algorithm is 0.0179 m, and the root-mean-square error of the original Cartographer algorithm is 0.0230 m. Compared with the original Cartographer algorithm, the root-mean-square error of the Iao_ICP algorithm is reduced by 22.06%. The average error of the Iao_ICP algorithm is 0.0044 m smaller than that of the original Cartographer algorithm. The maximum absolute trajectory error of the original Cartographer algorithm is 0.1428 m in this sequence, and the maximum absolute trajectory error of the Iao_ICP algorithm is 0.1357 m. The minimum absolute trajectory error of the Cartographer original algorithm is 0.0024 m, and the minimum absolute trajectory error of the Iao_ICP algorithm is 0.0011 m. It can be seen from the above data that the Iao_ICP algorithm has a smaller relative trajectory error than the original Cartographer algorithm in sequence ①.

6. Physical Experiment Analysis

This experiment uses a small wheeled differential car as the mobile robot platform.

As shown in Figure 11, the platform configuration is as follows: wheeled robot, embedded development board, 16-line RS-LIDAR-16 scanner, IPMS-IG IMU. Among them, the wheeled robot is driven by four wheels and two motors. The embedded development board uses STM32f103 as the main controller, and it is also equipped with a motor driver module and an MPU6050 module. RS-LiDAR-16 adopts a hybrid solid-state LiDAR, which integrates 16 laser transceiver components. The measurement distance is up to 150 m, the measurement accuracy is within ± 2 cm, the number of output points is up to 300,000 points/s, the horizontal angle is 360°, the vertical measurement is 360°, and the angle is $\pm 15°$. IMU integrates three-axis acceleration and angular velocity sensors, which can measure the real-time pose of the robot, and has the advantages of high precision, high frequency, low power consumption, and strong real-time performance. This experiment realizes the conversion of 3D LiDAR to 2D LiDAR by projecting the 16-line data of 3D LiDAR onto a fixed plane. Since the real motion trajectory of the robot cannot be accurately obtained in the real scene, this experiment judges and tests the cumulative error of the robot pose during the mapping process of the Iao_ICP algorithm according to the loopback effect. The movement of the robot is controlled by the handle in this experiment.

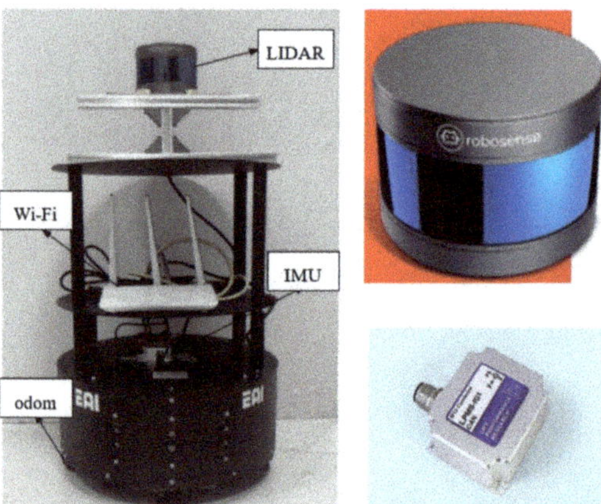

Figure 11. Mobile experiment platform.

The real environment is a rectangular hall corridor with a length of about 43 m, a width of about 51 m, and a building area of about 2193 m^2, as shown in Figure 12 above. It is easy to measure the actual size of the object and compare the data with the mapping accuracy of the test algorithm. Due to the cabinets, building supports, stair entrances, elevator entrances, and other objects in the environment have a strong structure, the effectiveness and robustness of the algorithm for eliminating laser motion distortion and mapping accuracy can be tested in the above environment. There are also the following reasons: the test scene is relatively large, and there are long straight corridors, transparent glass, flowing crowds, and other factors in the environment that may easily interfere with the test of mapping. The smooth marble floor increases the accumulation of pose errors during the movement of the robot.

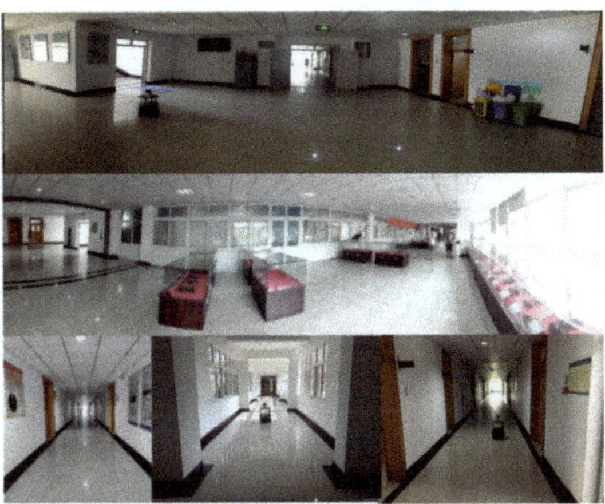

Figure 12. Experimental real scene.

To compare the mapping accuracy of the Iao_ICP algorithm and the original Cartographer algorithm, 10 highly structured objects were selected in the test scene for measurement and analysis. Figures 13 and 14 are the mapping effect of the original Cartographer algorithm and the mapping effect of the Iao_ICP algorithm. First, the actual size of the object is measured by a handheld laser rangefinder. The map measurements displayed in the rviz plugin for algorithmic mapping are measured. Finally, the relative error and absolute error of the two algorithms are calculated. The measurement data and error values of the above two algorithms are shown in Tables 2 and 3 below. Figure 15 is a comparison chart of the relative error of the two algorithms.

It can be seen from Figures 13 and 14 that the original Cartographer algorithm has a large pose error product in this experimental scene. Although a loop can be formed, the effect of eliminating local errors on the map is not good. The Iao_ICP algorithm removes motion distortion from most laser data by fusing wheel odometer and IMU information. At the same time, the laser scan data are compensated by estimating the speed of the robot and ICP algorithm. The Iao_ICP algorithm not only effectively removes motion distortion, but also eliminates the accumulation of pose errors caused by tire slippage during robot motion. Figure 14 shows that the map constructed by the Iao_ICP algorithm has no confusion, no burrs, and clear structural features. It can clearly express the surrounding environment information, and the map ghost is small. It can be seen that the mapping effect of the Iao_ICP algorithm is better than that of the original Cartographer algorithm. Combined with the error data analysis in Tables 2 and 3, and Figure 15, it can be seen that the average relative error of the Iao_ICP algorithm is much smaller than that of the original Cartographer algorithm, and the relative error is mostly concentrated below 1%. The error is stable, and there is no mutation.

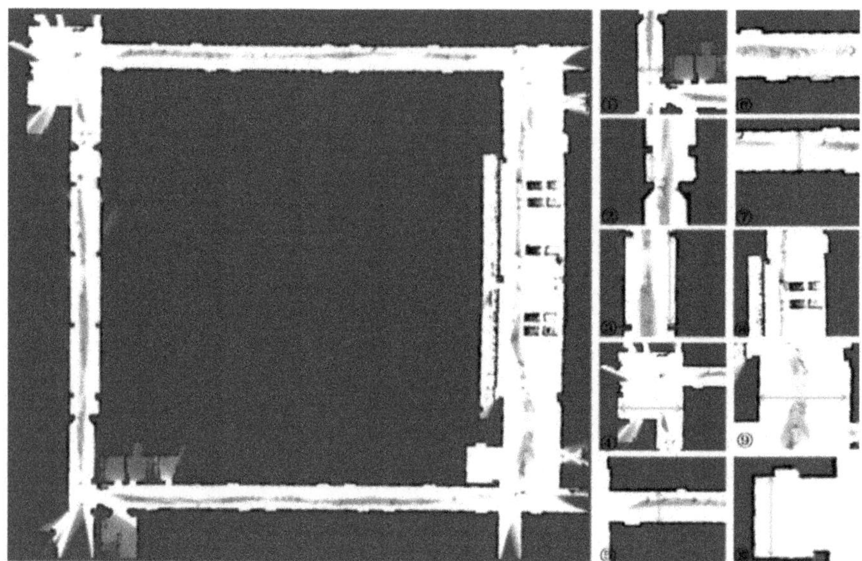

Figure 13. Mapping effect of Cartographer.

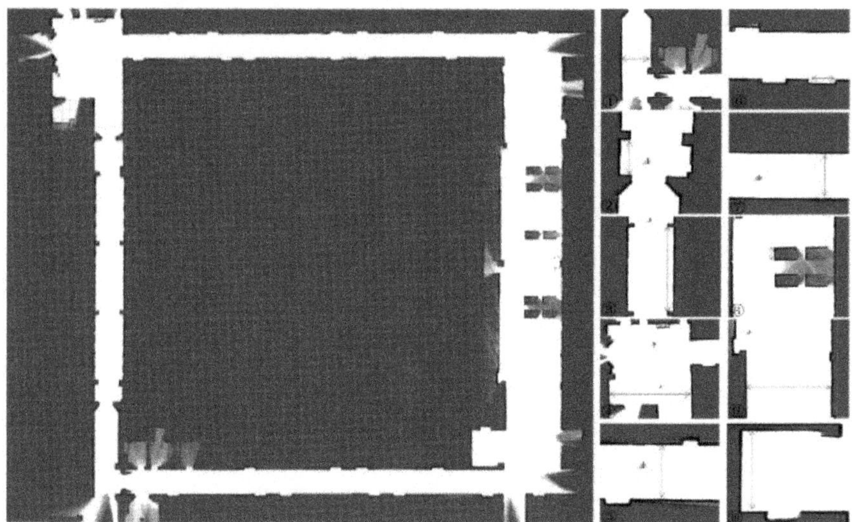

Figure 14. Mapping effect of Iao_ICP.

Table 2. Cartographer original algorithm mapping error table.

Measuring Point	Measured Value (cm)	Figure Measured Values (cm)	Absolute Error (cm)	Relative Error (%)
1	284.700	288.074	−3.374	−1.185107
2	195.000	186.400	8.600	4.410256
3	712.200	709.709	2.491	0.349761
4	812.000	803.200	8.800	1.083743
5	271.000	263.200	7.800	2.878228
6	136.300	130.840	5.460	4.005869
7	272.300	264.320	7.980	2.930591
8	76.500	85.895	−9.395	−12.281045
9	629.200	627.426	1.774	0.281945
10	402.700	397.230	5.470	1.358331

Table 3. Iao_ICP algorithm mapping error table.

Measuring Point	Measured Value (cm)	Figure Measured Values (cm)	Absolute Error (cm)	Relative Error (%)
1	284.700	283.700	1.000	0.351246
2	195.000	197.540	3.460	1.774358
3	712.200	712.363	−0.163	−0.022886
4	812.000	819.340	−7.340	−0.903940
5	271.000	270.928	0.072	0.026568
6	136.300	133.549	2.751	2.018341
7	272.300	270.116	2.184	0.802056
8	76.500	77.744	−1.244	−1.626143
9	629.200	626.829	2.371	0.376827
10	402.700	404.365	−1.665	−0.413459

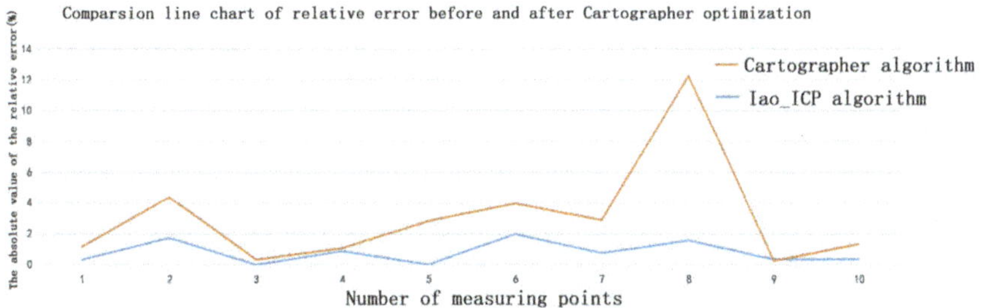

Figure 15. Line chart of relative error comparison of two algorithms.

7. Conclusions

For the problem of removing laser motion distortion, in the case of wheel slippage and accumulated error, the traditional method of directly measuring displacement and angle information based on the wheel odometer, and the odometer angle data obtained by the encoder, will have a certain deviation. In addition, with the traditional method of directly measuring the angular velocity and linear acceleration based on the inertial navigation unit, and then integrating the displacement and angle information, due to the poor accuracy of the linear acceleration of the IMU, the local accuracy of the quadratic integration is still very poor. Therefore, the displacement data obtained will also have a certain deviation. The Iao_ICP algorithm proposed in this paper uses the linear interpolation method to obtain the pose of the LiDAR, which solves the alignment problem of discontinuous laser scan data. Data fused by IMU and odometer provide a better initial value for ICP. The estimated speed is introduced as the termination condition of the ICP method iteration to realize the compensation of the LiDAR data. The experiment uses a small wheeled mobile robot to collect data and compare and analyze results in a corridor environment to verify the original Cartographer algorithm and the Iao_ICP algorithm. Finally, the experimental data show that the algorithm proposed in this paper can effectively remove laser motion distortion, improve the accuracy of mapping, and reduce the cumulative error.

Author Contributions: Conceptualization, Q.W. and Q.M.; methodology, Q.M.; software, Z.Z.; validation, Q.W., C.L. and P.Z.; formal analysis, Z.Z.; investigation, Q.M.; resources, Q.W.; data curation, B.Z.; writing—original draft preparation, Q.M.; writing—review and editing, Z.Z.; visualization, C.L.; supervision, Y.L.; project administration, W.M.; funding acquisition, Y.T. All authors have read and agreed to the published version of the manuscript.

Funding: This work was supported by Science and Technology Project of the State Grid Corporation of China "Research and application of visual and auditory active perception and collaborative cognition technology for smart grid operation and maintenance scenarios" (Grant: 5600-202046347A-0-0-00).

Institutional Review Board Statement: This study was conducted in accordance with the Declaration of Helsinki, and approved by the Ethics Committee of Hangzhou Dianzi University (protocol code CE022108 22/02/26).

Informed Consent Statement: Informed consent will be obtained from all subjects involved in this study.

Data Availability Statement: Data will be made available upon request from the authors.

Conflicts of Interest: The authors declare no conflict of interest.

References

1. Li, P.; Shengze, W.; Michael, K. π-SLAM:LiDAR Smoothing and Mapping With Planes. In Proceedings of the 2021 IEEE International Conference on Robotics and Automation, Xi'an, China, 19–22 November 2021; pp. 5751–5757.
2. Rui, H.; Yi, Z. High Adaptive LiDAR Simultaneous Localization and Mapping. *J. Univ. Electron. Sci. Technol. China* **2021**, *50*, 52–58.
3. Xin, L.I.; Xunyu, Z.; Xiafu, P.; Zhaohui, G.; Xungao, Z. Fast ICP-SLAM Method Based on Multi-resolution Search and Multi-density Point Cloud Matching. *Robot* **2020**, *42*, 583–594.
4. Hess, W.; Kohler, D.; Rapp, H.; Andor, D. Real-time loop closure in 2D LI-DAR SLAM. In Proceedings of the IEEE International Conference on Robotics and Automation (ICRA), Stockholm, Sweden, 16–21 May 2016.
5. David, J.Y.; Haowei, Z.; Mona, G.; Hugues, T.; Timothy, D.B. Unsupervised Learning of LiDAR Features for Use in a Probabilistic Trajectory Estimator. In Proceedings of the 2021 IEEE International Conference on Robotics and Automation, Xi'an, China, 19–22 November 2021; pp. 1154–1161.
6. Jo, J.H.; Moon, C.-b. Development of a Practical ICP Outlier Rejection Scheme for Graph-based SLAM Using a Range Finder. *Korean Soc. Precis. Eng.* **2019**, *20*, 1735–1745. [CrossRef]
7. Xue, H.; Fu, H.; Dai, B. IMU-aided high-frequency LiDAR odometry for autonomous driving. *Appl. Sci.* **2019**, *9*, 1506. [CrossRef]
8. Bezet, O.; Cherfaoui, V. Time error correction for range scanner data. In Proceedings of the International Conference on Information Fusion, New York, NY, USA, 10–13 July 2006.
9. Hong, S.; Ko, H.; Kim, J. VICP:velocity updating iterative closest point algorithm. In Proceedings of the IEEE International Conference on Robotics and Automation, Anchorage, AK, USA, 3–8 May 2010; pp. 1893–1898.
10. Zhang, J.; Singh, S. Low-drift and real-time LiDAR odometry and mapping. *Auton Robot.* **2017**, *41*, 401–406. [CrossRef]
11. Arun, K.S.; Huang, T.S.; Blostein, S.D. Least-squares fitting of two 3-D point sets. *IEEE Trans. Pattern Anal. Mach. Intell.* **1987**, *9*, 698–700. [CrossRef] [PubMed]
12. Xiaohui, J.; Wenfeng, X.; Jinyue, L.; Tiejun, L. Solving method of LiDAR odometry based on IMU. *J. Instrum.* **2021**, *42*, 39–48.
13. Xu, Z.; Zhi, W.; Can, C.; Zhixuan, W.; Renxin, H.; Jian, W. Design and Implementation of 2D LiDAR Positioning and Mapping System. *Opt. Technol.* **2019**, *45*, 596–600.
14. Yokozuka, M.; Koide, K.; Oishi1, S.; Banno, A. LiTAMIN2: Ultra Light LiDAR-based SLAM using Geometric Approximation applied with KL-Divergence. In Proceedings of the 2021 IEEE International Conference on Robotics and Automation, Xi'an, China, 19–22 November 2021; pp. 11619–11625.
15. Liwei, L.; Xukang, Z.; Xiuhua, L.; Zijun, Z. Research on Map Evaluation of 2D SLAM Algorithm for Low-Cost Mobile Robots. *Comput. Simul.* **2021**, *38*, 291–295.

Article

A Hybrid Prognostic Method for Proton-Exchange-Membrane Fuel Cell with Decomposition Forecasting Framework Based on AEKF and LSTM

Zetao Xia [1], Yining Wang [1], Longhua Ma [2], Yang Zhu [3], Yongjie Li [2], Jili Tao [2] and Guanzhong Tian [1,*]

[1] Ningbo Innovation Center, Zhejiang University, Ningbo 315000, China
[2] School of Information Science and Engineering, NingboTech University, Ningbo 315000, China
[3] College of Control Science and Engineering, Zhejiang University, Hangzhou 310027, China
* Correspondence: gztian@zju.edu.cn

Abstract: Durability and reliability are the major bottlenecks of the proton-exchange-membrane fuel cell (PEMFC) for large-scale commercial deployment. With the help of prognostic approaches, we can reduce its maintenance cost and maximize its lifetime. This paper proposes a hybrid prognostic method for PEMFCs based on a decomposition forecasting framework. Firstly, the original voltage data is decomposed into the calendar aging part and the reversible aging part based on locally weighted regression (LOESS). Then, we apply an adaptive extended Kalman filter (AEKF) and long short-term memory (LSTM) neural network to predict those two components, respectively. Three-dimensional aging factors are introduced in the physical aging model to capture the overall aging trend better. We utilize the automatic machine-learning method based on the genetic algorithm to train the LSTM model more efficiently and improve prediction accuracy. The aging voltage is derived from the sum of the two predicted voltage components, and we can further realize the remaining useful life estimation. Experimental results show that the proposed hybrid prognostic method can realize an accurate long-term voltage-degradation prediction and outperform the single model-based method or data-based method.

Keywords: prognostics; proton-exchange-membrane fuel cell; hybrid method; degradation prediction; remaining useful life

Citation: Xia, Z.; Wang, Y.; Ma, L.; Zhu, Y.; Li, Y.; Tao, J.; Tian, G. A Hybrid Prognostic Method for Proton-Exchange-Membrane Fuel Cell with Decomposition Forecasting Framework Based on AEKF and LSTM. *Sensors* **2023**, *23*, 166. https://doi.org/10.3390/s23010166

Academic Editor: Chun Sing Lai

Received: 20 October 2022
Revised: 30 November 2022
Accepted: 14 December 2022
Published: 24 December 2022

1. Introduction

Owing to the global energy crisis and environmental pollution that humans face, fuel-cell technology has attracted more and more attention from researchers as well as commercial companies. With the advantages of clean, high energy efficiency, and low operating temperature [1,2], the proton-exchange-membrane fuel cell (PEMFC) has been considered as one of the most attractive energy devices for future power applications. However, the durability and the high cost of PEMFC have been the bottlenecks of its large-scale commercial deployment. During operation, the components of the fuel cell, including the proton-exchange-membrane (PEM), the bipolar plate, the gas diffusion layer (GDL), the catalyst layer, and a membrane will degrade due to different working conditions and load cycling [3]. The performance of PEMFC suffers from multiple failure mechanisms, such as conductivity loss, catalyst reaction activity, and mass transfer [4]. The performance of a PEMFC system is characterized by its efficiency and cyclability, which are highly influenced by membrane properties [5]. Shanmugam et al. [6] developed a new block copolymer membrane with a lower self-discharge rate. The cyclability with slight capacity decay showed its chemical stability for long-term operation. Rajput et al. [7] synthesized a graphene oxide composite membrane which has better mechanical and thermal stability. Furthermore, working under highly dynamic conditions, especially in automotive applications, will accelerate the aging process of PEMFC and increase the probability of failure occurrence [8]. The International Energy Agency reported that the cost of commercial fuel cell stack is less than 10,000 USD/kW in 2017 [9]. The maintenance costs drastically decreased from 40 EUR Ct/kWh in 2012 to 20 EUR Ct/kWh in 2017.

The balance of plant (BOP), which mainly consists of an air-supply system, hydrogen-circulation system, water-and-heat-management system and control system, maintains the stable and safe operation of the stack [10]. The degradation mechanism is too complicated to be fully understood with the current technology. To extend the fuel-cell lifetime and reduce its maintenance cost, the management and control strategy of the PEMFC has become a hot research topic. The prognostic method provides a potential solution to extending the PEMFC lifespan [11,12]. As the prerequisite for the maintenance of PEMFC, an effective prognostic method can estimate the state of health (SOH) of the fuel cell and predict the system's future evolution. By prognostic methods, the degradation process of PEMFC can be investigated and modeled [13] to guide the maintenance services of fuel cells before failures occur. The prognostic methods of PEMFC can generally be divided into three categories [2,14]: the model-based method, the data-based method, and the hybrid method.

The model-based method uses the mechanism degradation model or the empirical degradation model to realize the prognostics of PEMFC. The mechanism model adopts mathematical equations to describe the internal aging process, with the advantages of less training data and strong generality. However, it suffers from a large computational burden and a high complexity of the degradation mechanism [4,14]. Zhang and Pisu [15] built the catalyst degradation model to describe the relationship between operating conditions and the degradation rate of electrochemical surface area (ECSA). Dhanushkodi et al. [16] developed a diagnostic method to characterize the catalyst component durability. Based on the Pt/C catalyst degradation mechanism [17], Polverino and Pianese [18] proposed the dissolution-mechanism model and the Ostwald-ripening-mechanism model to estimate ECSA. However, the validity of the mechanism model needs to be verified by the experimental data, and the adjustment of model parameters depends on expert experience. The empirical degradation model with less computational burden is easier to deploy in online applications. Jouin et al. [19] proposed a PEMFC prognostic method based on logarithmic, polynomial, and exponential empirical equations. Bressel et al. [13] proposed a typical semi-empirical prognostic method that brings polarization curves into consideration. Li et al. [20] proposed an estimation algorithm for lithium-battery SOC in electric vehicles based on an adaptive unscented Kalman filter (AUKF). Zhang et al. [1] realized internal characterization-based prognostics for fuel cells based on a Markov-process algorithm.

Data-based methods can be conducted without considering the complex mechanism of PEMFC and can improve the prediction accuracy as long as sufficient monitoring data are available. Silva et al. [21] developed a long-term prediction model for PEMFC based on the adaptive neuro-fuzzy inference system (ANFIS). The wavelet decomposition is proposed in [22] to improve short-term prediction accuracy. In [23,24], the echo state network (ESN) is adopted for forecasting the degradation process. Ma et al. [25] adopted a long short-term memory network (LSTM) to predict the degradation voltage, which identified the superiority of the LSTM network compared with the relevance vector machine (RVM) and the Elman network. Yang et al. [12] proposed an RUL prediction method for the bearing's degradation process based on LSTM. However, the data-based method suffers from poor generality in practical deployment and there is a shortage of training data because of the costly and time-consuming PEMFC aging test.

The hybrid method is established by combining the advantages of the model-based method and the data-based method through different strategies [26]. It is usually more accurate and robust than a single method at the cost of a more complex structure and a higher computational burden [2]. Peng et al. [11] realized the RUL estimation for a turbofan engine based on the convolutional neural networks (CNN) and long short-term memory (LSTM) structures. Li et al. [27] used a linear-parameter-varying model to build the virtual stack voltage as the health indicator and the degradation trend was predicted by ensemble ESN. Ma et al. [28] fused the extended Kalman filter (EKF) and

LSTM algorithms to realize a more accurate prediction result. EKF is used to estimate the system state and then the prediction of LSTM is regarded as the observation for EKF. Based on ANFIS [21], Liu et al. [26] realized the long-term degradation trend prediction and the remaining useful life (RUL) estimation is achieved by AUKF. The membership function is optimized automatically by a particle swarm optimization (PSO) algorithm. The methods above mainly focus on developing new prediction structures. However, the long-term prediction accuracy of those methods still cannot meet expectations: the prediction effectiveness under automotive load cycling needs to be improved.

The voltage-recovery phenomenon occurs periodically after the characterization test and it significantly influences the prediction accuracy. The investigation into this phenomenon can reveal the aging process of fuel cells and support appropriate maintenance strategies. Jouin et al. [29] combined the global power-aging model and power-recovery model based on the particle filter (PF) algorithm to forecast voltage degradation. Morando et al. [30] used the wavelet filter to decompose the stack voltage into two parts and make predictions based on ESN. With the introduction of the self-healing factor, Kimotho et al. [31] realized the prediction of the voltage-aging process after each characterization. Deng et al. [32] proposd a novel empirical model based on the PF algorithm for the remaining useful-life prediction of a lithium-ion battery. The authors separated the local degradation process from the global degradation process to capture the degradation and regeneration phenomena. Zhou et al. [33] divided the voltage data into stationary and non-stationary sequences. Then, the autoregressive and moving average (ARMA) model and time-delay neural network (TDNN) were utilized to predict the degradation voltage. However, the prediction of those models is not robust or accurate enough, as the voltage-recovery phenomenon possesses strong nonlinearity.

Since the PEMFC degradation-process mechanism has not been fully investigated yet, the model-based method's prediction accuracy cannot meet expectations. The data-based method cannot give a satisfying prediction with enough long-term forecasting horizon. Moreover, the voltage-recovery phenomenon is still a problem for most of the prognostic methods. Thus, it is of great significance to explore a hybrid method to combine the advantages of those two methods to better predict the PMEFC degradation process. In addition, the parameter-adjustment process requires a lot of manual intervention which is very time-consuming. Therefore, it is meaningful to realize model construction and hyperparameters optimization automatically.

A hybrid prognostic method for PEMFC based on the decomposition forecasting framework is proposed in this paper. Specifically, the original voltage data is decomposed into the calendar aging components and the reversible aging components based on the locally weighted regression method (LOESS). Then, we apply the calendar aging model based on an adaptive extended Kalman filter (AEKF) and the reversible aging model based on LSTM to predict the two voltage components, respectively. In this way, the aging process of the PEMFC, including the voltage-recovery phenomenon, can be better forecasted. The final predicted voltage is derived from the sum of the two predictions, and we can further realize RUL estimation. The main contributions of this paper are summarized as follows:

(1) We establish the decomposition forecasting framework to predict the long-term voltage degradation of PEMFC. After the decomposition by LOESS, we apply the AEKF algorithm and the LSTM neural network to predict those two components, respectively. This framework can combine the AEKF method's advantage of predicting overall aging trends and the LSTM model's advantage of strong nonlinear-modeling ability. An iterative structure is adopted to realize the long-term degradation voltage forecasting.

(2) Based on the physical aging model, we develop three-dimensional aging factors to better characterize the fuel cell's aging state. Considering the voltage-recovery phenomenon, we adopt a sliding-window strategy during the training of the LSTM network to improve the prediction accuracy of the model.

(3)   The automatic machine-learning (AutoML) method based on the genetic algorithm is adopted to optimize the hyperparameters of the LSTM network automatically, which can improve the prediction accuracy and training efficiency.

The remaining contents of this paper are organized as follows. In Section 2, the decomposition forecasting framework is introduced, followed by the configurations of the AEKF model and the LSTM network. The prediction results and discussions of our method are presented in Section 3. Finally, the conclusion is summarized in Section 4.

2. Methodology

2.1. The Decomposition Forecasting Framework

The framework of the proposed hybrid prognostic method for PEMFC is shown in Figure 1. First of all, the original voltage data were decomposed into the calendar aging part and the reversible aging part by LOESS. Then, we established the calendar aging model based on the AEKF algorithm to predict the overall aging trend for PEMFC. The genetic algorithm was applied to identify the parameters of physical aging model from the polarization curve. The three-dimensional aging factors were introduced in physical aging model to better depict the degradation trend. Next, based on the LSTM network, we built the reversible aging model to capture the voltage-recovery information. AutoML approach was adopted in the training phase of LSTM for the hyperparameters tuning automatically. In addition, the iterative structure was utilized to realize long-term degradation forecasting [30]. The final prediction of the aging voltage can be obtained by combining the two predicted components and we can further realize RUL estimation.

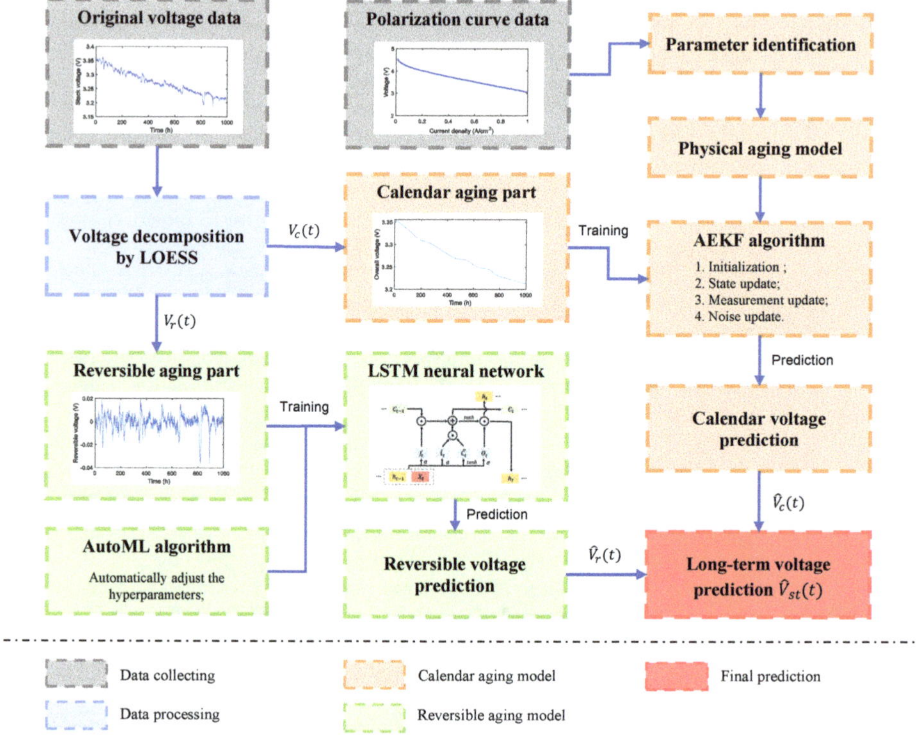

Figure 1. The decomposition forecasting framework of the proposed hybrid method.

2.2. Dataset Analysis

The dataset we used in this paper comes from IEEE PHM 2014 Data Challenge [34], conducted and collected by FCLAB. The FC1 has a constant current of 70 A while 10% triangular current ripples with the frequency of 5 kHz are added to the 70 A current for FC2. The monitoring data were obtained during the aging test, including voltage, operating parameters, electrochemical impedance spectroscopy (EIS) measurement, and polarization curve. The test bench was adapted for 1 kW fuel cell stack. To master the fuel cells' running conditions accurately, the experimental operating parameters of the PEMFC can be regulated and measured as shown in Table 1. The gas-humidification subsystem is composed of the two boilers placed upon the stack. Air and hydrogen flow through respective boilers before reaching the stack. Only the air boiler is heated to obtain the required relative humidity. The hydrogen boiler is kept at room temperature due to the need for dry anode gas. The cooling water subsystem dominates the temperature of the stack. The stack voltage is selected as the health indicator of PEMFC degradation since it can be measured easily and it is suitable for online applications [33]. Since the degradation process of fuel cells is slow, the dataset was down-sampled with the interval of one hour to reduce the computational burden. Each considered fuel-cell stack consisted of five cells. The length of FC 1 and FC 2 are 991 h and 1020 h, respectively.

Table 1. PEMFC stack and experimental operating parameters.

Parameter	Control Range
Number of cells	5
Active area	$100\ cm^2$
Load current	70 A (FC1)/63–77 A (FC2)
Operating hours	991 h (FC1)/1020 h (FC2)
Air flow rate	23 L/min
Hydrogen flow rate	4.8 L/min
Coolant flow rate	2 L/min
Pressure of anode and cathode	1.3 bar
Stack temperature	55 °C
Relative humidity	50%

In Figure 2, it is easy to see that the voltage always increases after the characterization test, which is marked by black circles. This is the voltage-recovery phenomenon mainly caused by the interruption of continuous testing during the rest periods [29]. During this time, the water content and distribution of the catalysts return to the previous state, which contributes to ECSA and the proton transfer. The interruption time for characterizations is scheduled weekly, at about 48 h, 185 h, 348 h, 515 h, 658 h, and 823 h for FC1 and 35 h, 182 h, 343 h, 515 h, 666 h, and 830 h for FC2. In addition, it can be noticed in Figure 2 that sudden voltage drops occurred in the dashed boxes, which are regarded as faults during the aging test.

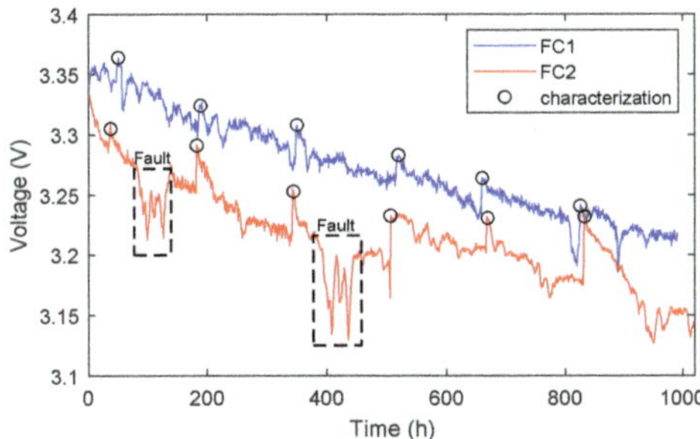

Figure 2. The voltage-degradation curves of FC1 and FC2.

2.3. Voltage Decomposition

Locally Weighted Regression

Motivated by the idea of decomposition forecasting, the original voltage data is decomposed into the calendar aging part and the reversible aging part by LOESS. LOESS is a nonparametric method for regional regression analysis, which mainly divides the samples into small windows and performs polynomial fitting on them. Repeating this process continuously, we can finally obtain the regression curve. The points near the fitting point have a greater impact on the regression curve and the weight is constructed by the tricube weight function [35]. The weight of fitting points is defined as follows:

$$w_i = \begin{cases} \left(1 - \left|\frac{x-x_i}{\Delta(x)}\right|^3\right)^3, \left|\frac{x-x_i}{\Delta(x)}\right| \leq 1 \\ 0, \left|\frac{x-x_i}{\Delta(x)}\right| \geq 1 \end{cases} \tag{1}$$

where $\Delta(x)$ is the size of the window, x_i is the fitting point, and x is the center of the window. The weighted regression can be carried out based on the weighted least-square method.

After the original voltage $V_{st}(t)$ was filtered by LOESS, we could obtain the calendar aging component $V_c(t)$. Then, the original voltage $V_{st}(t)$ was subtracted from $V_c(t)$ to obtain the reversible aging component $V_r(t)$. Thus, the fuel-cell stack voltage can be divided into two parts:

$$V_{st}(t) = V_c(t) + V_r(t) \tag{2}$$

As shown in Figure 3, we adopted an iterative structure to realize the long-term time series forecasting [30]. When forecasting h steps ahead, we used the value $\hat{y}_{k+1}$ just forecasted by a one-step prediction model as part of the input variables for forecasting the next step, where u_k represents the input. We continued in this manner until the desired prediction horizon was reached. In particular, prediction errors accumulated through this strategy, which may lead to a divergence in results [36].

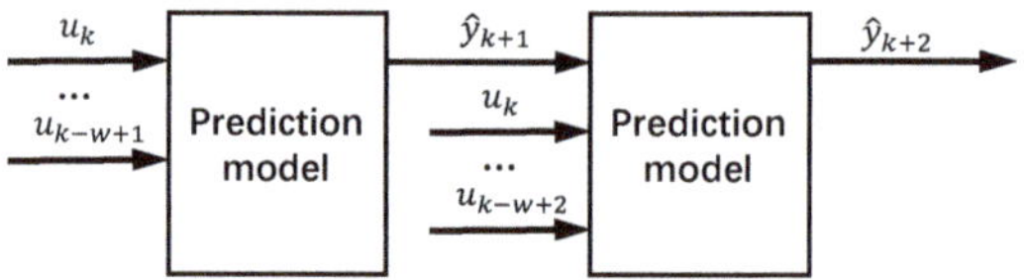

Figure 3. Iterative structure.

2.4. Calendar Aging Model Based on AEKF

2.4.1. Physical Aging Model

Previous studies have shown that the polarization curve changes regularly as the operation of PEMFC continues [37], which enables us to build a degradation model based on it. The empirical model of the polarization curve introduced in [13] can be expressed as Equation (3).

$$V_c(t) = N\left(E_{ocv} - i(t)R - aT\ln\left(\frac{i(t)}{i_0}\right)\right.$$
$$\left. + bT\ln\left(1 - \frac{i(t)}{i_L}\right)\right) \tag{3}$$

where V_c is the calendar aging voltage, which represents the approximate part of the stack voltage, N is the number of cells, i is the stack current, T is the operation temperature, a is the Tafel constant, b is the concentration constant, E_{ocv} is the open-circuit voltage, R is the total resistance, i_0 is the exchange current, and i_L is the limiting current.

According to the study in [13], only R and i_L vary with the operating time obviously during the aging test. Parameters E_{ocv} and i_0 changed a little, so they can be assumed as constant values. The increase in R may result from the polymer membrane's degradation and the plates' corrosion [38]. The decrease in i_L is related to the ripening of the platinum particles and poor hydrophobicity of GDL, which accounts for the reduction in the mass transfer [39]. Therefore, an aging factor α is introduced to describe the change in the aging parameters (R and i_L), since they have similar change speeds [13,26]. The empirical aging parameters can be expressed by Equation (4):

$$\begin{cases} R = R_0(1 + \alpha(t)) \\ i_L = i_{L0}(1 - \alpha(t)) \\ \alpha(t) = \beta t, \beta(t) = \gamma t \end{cases} \tag{4}$$

where $\alpha(t)$ represents the degradation state of the fuel cell, $\beta(t)$ represents the fuel-cell degradation rate, and $\gamma(t)$ is the derivative of $\beta(t)$. We notice that a constant $\beta(t)$ will lead to a linear change in the degradation state $\alpha(t)$, which will reduce the prediction accuracy of the model. Therefore, we introduced another factor, $\gamma(t)$, so that the degradation rate $\beta(t)$ can change with time to better forecast the variation in the aging trend. As a result, the three-dimensional aging factors consist of $\alpha(t)$, $\beta(t)$, and $\gamma(t)$.

Combining Equations (3) and (4), we can obtain the expression of the physical aging model as follows:

$$\begin{aligned} V_c(t) = \ & N\left(E_{ocv} - R_0(1 + \alpha(t))i(t) - aT\ln\left(\frac{i(t)}{i_0}\right) \right. \\ & \left. + bT\ln\left(1 - \frac{i(t)}{i_{L0}(1 - \alpha(t))}\right) \right) \end{aligned} \tag{5}$$

The degradation process of a fuel cell is nonlinear and can be expressed as Equation (6):

$$\begin{cases} x_k = f(x_{k-1}) + w_{k-1} \\ y_k = g(x_k, u_k) + v_k \end{cases} \tag{6}$$

where x_k is the aging state at kth sampling time, u_{k-1} is the input(current), y_k is the system output (stack voltage), w_k and v_k represent the process and measurement noises which are assumed to obey Gaussian distribution with zero mean and variances of Q and R, and $f(\cdot)$ and $g(\cdot)$ are functions used to describe the degradation model.

To better forecast the aging trend of PEMFC, here we introduce three-dimensional aging factors which can be expressed as Equation (7):

$$x_k = [\alpha_k, \beta_k, \gamma_k]^T \tag{7}$$

where α_k is the value of degradation state at kth sampling time, β_k is the degradation rate at kth sampling time, and γ_k is the derivative of β. Since $u_k = i_k$, $y_k = V_{c,k}$, the discrete time-state-space equation for PEMFC can be expressed as follows:

$$\begin{cases} \begin{bmatrix} \alpha_k \\ \beta_k \\ \gamma_k \end{bmatrix} = \begin{bmatrix} 1 & \Delta t & 0 \\ 0 & 1 & \Delta t \\ 0 & 0 & 1 \end{bmatrix} \begin{bmatrix} \alpha_{k-1} \\ \beta_{k-1} \\ \gamma_{k-1} \end{bmatrix} + w_{k-1} \\ y_k = N \cdot \left[E_{ocv} - R_0(1 + \alpha_k)i_k - aT\ln\left(\frac{i_k}{i_0}\right) \right. \\ \qquad\qquad \left. + bT\ln\left(1 - \frac{i_k}{i_{L0}(1 - \alpha_k)}\right) \right] + v_k \end{cases} \tag{8}$$

where Δt represents the sample period. Here, the parameters, including E_{ocv}, R_0, a, b, i_0, and i_{L0}, need to be identified to initialize our calendar aging model.

In order to avoid overfitting, Akaike information criterion (AIC) can be used to measure the fitting results of the proposed model [40]. In general, AIC can be expressed as:

$$AIC = 2k - 2\ln(L) \tag{9}$$

where k is the number of parameters and L is the likelihood function.

Let n be the number of observations and SSR represent the sum of the squares of the residuals; then, AIC becomes:

$$AIC = 2k + n\ln(SSR/n) \tag{10}$$

$$SSR = \sum (y_i - \hat{y}_i)^2 \tag{11}$$

AIC criterion is used to judge the goodness and the efficiency of the degradation models with two and three parameters (i.e., degradation state, its first derivative, and its second derivative). It can be seen from the Table 2 that AIC of the degradation model with three parameters is less than that of the model with two parameters. The smaller the AIC value, the better the model performance. Therefore, we chose three parameters to build our degradation model.

We regard the mean value of the state estimation as the optimal state estimate, and the point estimation of the trend components can be calculated by the system output matrix. Therefore, we can combine it with the prediction result of LSTM to obtain the final voltage prediction.

Table 2. AIC of the degradation models with different parameters for FC1 and FC2.

Stack	Training Data	Numbers of Parameters	
		2	**3**
FC1	55%	−3677	−4257
	70%	−2687	−2774
	80%	−1792	−1818
FC2	55%	−3564	−3845
	70%	−2347	−2479
	80%	−1545	−1622

2.4.2. Parameter Identification

We identified the parameters of our calendar aging model from the polarization curve data. Considering the multi-parameters and nonlinearity of the physical aging model, we chose the genetic algorithm to realize the parameter identification [39]. The aging factor α_k remains at zero since the polarization curve was measured at the beginning of the operation.

The genetic algorithm first initializes the values randomly, and then it performs selection, crossover, and mutation operations on individuals [41] according to the fitness function $f_{fitness}$. The optimal solution can be obtained through the iteration of the algorithm. The fitness function can be expressed as follows:

$$f_{fitness}\left(E_{ocv}, R_0, a, b, i_0, i_{L0}\right) = \sum_k \left[V_{c,k} - \hat{V}_{c,k}\right]^2 \tag{12}$$

where $V_{c,k}$ is the observed voltage and $\hat{V}_{c,k}$ is the estimated voltage. E_{ocv}, R_0, a, b, i_0, and i_{L0} are the parameters that need to be identified.

2.4.3. Extend Kalman Filter

In this paper, we applied the AEKF algorithm to deal with the nonlinearity of the fuel-cell system and to predict the calendar aging voltage. The traditional Kalman-filter algorithm assumes the process noise and the measurement noise as Gaussian white noise with zero means. However, it is difficult to obtain the statistical characteristics of noise in practice. Therefore, the AEKF method is introduced to correct the variance in those noises adaptively, to reduce the impact of unknown noise [42].

For the iterative calculation of our model, the Jacobian matrix can be obtained by linearizing the system with the first-order Taylor formula [39], as follows:

$$
\begin{cases}
A = \dfrac{\partial f(x_{k-1})}{\partial x}\bigg|_{x=\hat{x}_{k-1}^{+}} \\[2ex]
C_k = \dfrac{\partial g(x_k, u_k)}{\partial x}\bigg|_{x=\hat{x}_k^{-}}
\end{cases}
\tag{13}
$$

The algorithm of the discrete adaptive extended Kalman filter consists of four steps: initialization, state update, measurement update, and noise update, which are shown as follows:

1. Initialization: $\hat{x}_0^{+} = E[x_0]$, $P_0^{+} = E\left[\left(x_0 - \hat{x}_0^{+}\right)\left(x_0 - \hat{x}_0^{+}\right)^{T}\right]$, where $E[\cdot]$ is the mathematic expectation.
2. State update: $\hat{x}_k^{-} = A\hat{x}_{k-1}^{+} + w_{k-1}$, $P_k^{-} = AP_{k-1}^{+}A^{T} + Q_w$, where $\hat{x}_k^{-}$ is a-priori state estimate at step k, and P_k^{-} is a priori estimate error covariance.
3. Measurement update: $L_k = P_k^{-}C_k^{T}\left(C_k P_k^{-} C_k^{T} + R\right)^{-1}$, $\hat{x}_k^{+} = \hat{x}_k^{-} + L_k\varepsilon_k$, $P_k^{+} = (I - L_k C_k)P_k^{-}$, where $\varepsilon_k = y_k - g\left(\hat{x}_k^{-}, u_k\right)$, L_k is the Kalman gain at step k, $\hat{x}_k^{+}$ is a posteriori state estimate at step k, P_k^{+} is a posteriori estimate error covariance at step k.
4. Noise update: $\Pi_k = \frac{1}{M}\sum_{i=k-M+1}^{k}\varepsilon_i\varepsilon_i^{T}$, $Q_w = L_k\Pi_k L_k^{T}$, $R_v = \Pi_k - C_k P_k^{-} C_k^{T}$, where Π_k represents the mapping variance in error, and M represents averaging moving window of size.

2.5. Reversible Aging Model Based on LSTM

Long Short-Term Memory Networks

Through previous voltage decomposition, we can obtain the reversible aging voltage V_r which is the time-series sequence. The recurrent neural network (RNN) has a strong non-linear modeling ability for time-series data, which has achieved great success and wide application in natural language processing (NLP) [43] and time-series problems [44]. With the novel construction of the input gate, the forget gate, and the output gate, the LSTM network can overcome the problem of gradient disappearance or explosion from which traditional RNN suffers [25,35]. The LSTM network is applied to capture the voltage recovery information based on the reversible aging components in this paper. Figure 4a,b illustrates the LSTM architecture and the single cell of LSTM, respectively.

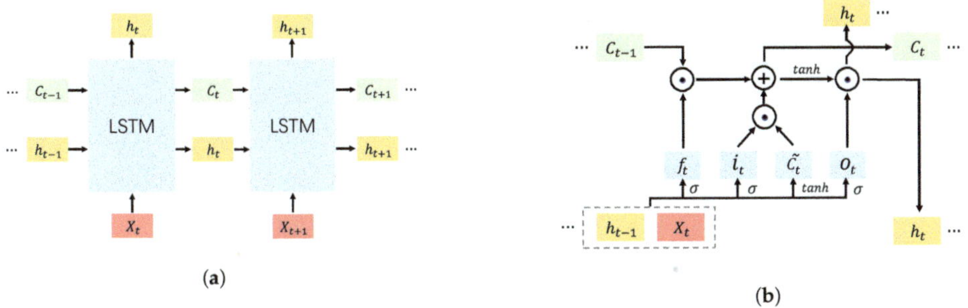

Figure 4. (**a**) LSTM architecture. (**b**) The single cell of LSTM.

Every time step, the LSTM unit receives the input from the current state X_t and the previous hidden state h_{t-1}, as Figure 4b shows. The expression of the input gate can be written as:

$$
i_t = \sigma(W_{xi}x_i + W_{hi}h_{t-1} + b_i)
\tag{14}
$$

The forget gate f_t determines which input information should be ignored from the history memory and it is defined as:

$$f_t = \sigma\left(W_{xf}x_t + W_{hf}h_{t-1} + b_f\right) \tag{15}$$

Meanwhile, the candidate value of the memory state $\tilde{C}_t$ is defined as:

$$\tilde{C}_t = \tanh(W_{xc}x_t + W_{hc}h_{t-1} + b_c) \tag{16}$$

Combining Equations (14)–(16), we can obtain the expression to update the cell state:

$$C_t = f_t \odot C_{t-1} + i_t \odot \tilde{C}_t \tag{17}$$

The output gate o_t is responsible for the final output and it is used to update the hidden state h_t based on the current cell state C_t. They can be written as follows:

$$o_t = \sigma(W_{xo}x_t + W_{ho}h_{t-1} + b_o) \tag{18}$$

$$h_t = o_t \odot \tanh(C_t) \tag{19}$$

where σ is the activation function and we choose the sigmoid function, W_{xi}, W_{hi}, W_{xf}, W_{hf}, W_{xc}, W_{hc}, W_{xo}, and W_{ho} are the weight matrices of each gate, b_i, b_f, b_c, and b_o are the bias vectors, $\odot$ means multiplied by the elements.

The residual components of the voltage data were smoothed by LOESS algorithm again with a window size of 20 to remove random noise or spikes before being sent to LSTM network. After smoothing to remove the noise, the reversible aging voltage data of PEMFCs and time information of characterization tests were input into the LSTM network as features. The network structure consists of four parts: a sequence input layer, an LSTM layer with the maximum number of 300 neurons in the hidden layer, a fully connected layer with one response, and a regression layer. The maximum sliding window size is 300; the loss function is the RMSE; the optimizer is Adam; the epoch size is 200; and the initial learning rate is 0.005.

We used the reversible voltage data and the time information of characterization from FC1 and FC2 to build samples for our training process of the network. We selected 50%, 70%, and 80% of the sample data as the training set, and the rest was selected as the test data set. The network's output is the reversible voltage at the next time step. Moreover, as shown in Figure 5, we adopted a sliding-window strategy during the training process of the LSTM. By setting the sliding window size reasonably, we can use the information from multiple times together as the feature input of the LSTM to improve the model's prediction ability for time-series data.

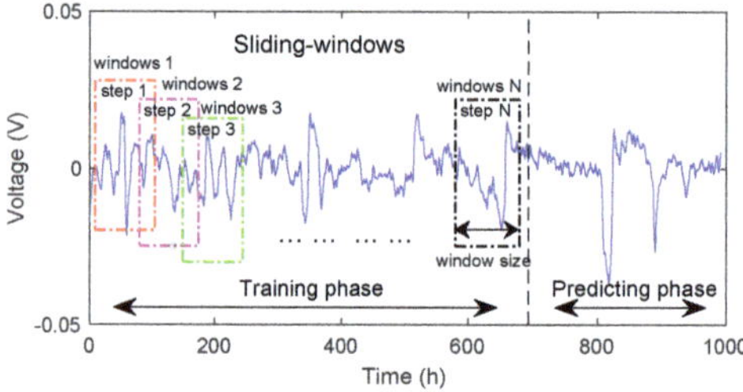

Figure 5. Sliding-window strategy during LSTM training.

2.6. AutoML Algorithm

The training of neural networks requires a lot of manual intervention which is very time-consuming. Here, with the help of the AutoML algorithm, we can realize model construction and hyperparameters optimization efficiently. Particularly, we apploed the genetic algorithm for finding appropriate hyperparameters of the LSTM network. The genetic algorithm is an optimization method inspired by the evolution process in nature selection [41].

The hyperparameters, including epochs, the number of neurons, and the sliding window size, were initialized arbitrarily ranging from 50 to 400, 50 to 300, and 10 to 300 with different intervals, respectively. Each individual in the population represents a potential solution to the problems to be resolved. The RMSE of LSTM prediction results in the test data set were used as its fitness value. Operations on individuals, including selection, crossover, and mutation, were performed to optimize the population. The parameters of genetic algorithm were set as follows: the population size is 12, the mutation rate is 0.2, the crossover rate is 0.5, and the iteration number is 5. The training process of LSTM was repeated until the genetic algorithm reached its maximum iteration.

3. Results and Discussions

The prediction results of our hybrid prognostics method are given in this section. Firstly, the voltage decomposition result is provided and then we will give and discuss our calendar aging model for PEMFC. Combined with the reversible aging model, we will obtain the final prediction.

3.1. Voltage Decomposition

The original voltage data of FC1 and FC2 are decomposed into the calendar aging part and the reversible aging part based on the LOESS, as shown in Figure 6. The smooth window size of FC1 is 300 and since FC2 fluctuates more violently, we set it to 500. The characterization tests (including the polarization curve test and the EIS measurement) are performed once a week, so the voltage recovery phenomenon appears periodically as shown in Figure 6c,d, where the red line indicates whether the characterization test was carried out. We can notice that the degradation of FC2 proves to be faster and more serious than FC1 because of its severe operating conditions.

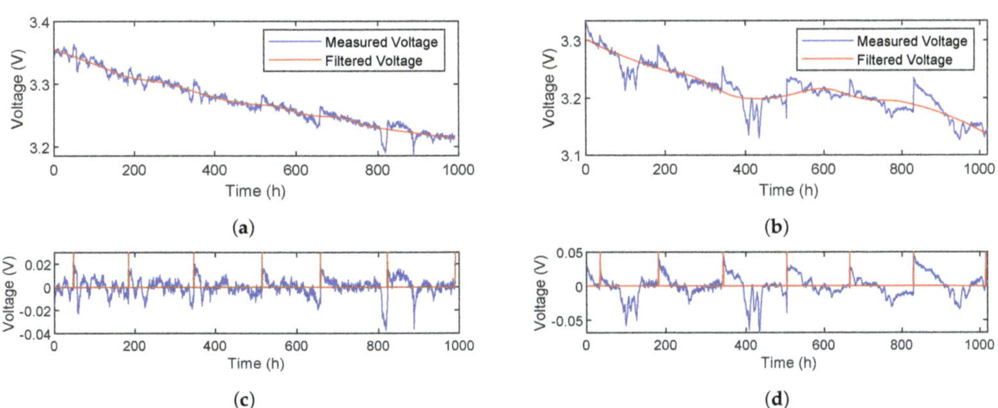

Figure 6. Voltage decomposition result. (**a**) Calendar aging voltage of FC1; (**b**) calendar aging voltage of FC2; (**c**) reversible aging voltage of FC1; (**d**) reversible aging voltage of FC2.

3.2. Calendar Aging Voltage Prediction

Here, we implement the calendar aging model based on the AEKF with the introduction of three-dimensional aging factors (T-AEKF) to better forecast the aging trend for PEMFC. The initial values of the state x_0, covariance matrix P_0, process error covariance matrix Q, and measurement error covariance matrix R were set as follows:

$$\begin{cases} x_0 = [8e\!-\!2, 2e\!-\!4, 2e\!-\!8]^T \\ P_0 = [0.1, 0, 0; 0, 0.01, 0; 0, 0, 0.0001] \\ Q = [5e\!-\!5, 0, 0; 0, 5e\!-\!5, 0; 0, 0, 5e\!-\!5] \\ R = 100 \end{cases}$$

Figure 7a,c,e shows the prediction results based on the T-AEKF for FC1 with 55%, 70%, and 80% training data, respectively. Figure 7b,d,f shows the prediction results based on the T-AEKF for FC2 with 55%, 70%, and 80% training data, respectively. The average values of the aging factors in the training phase were used for the iterative calculation in the predicting phase. The blue lines and the red dotted lines stand for AEKF output values in training and predicting phases, respectively. In Figure 7b,d,f the T-AEKF prediction result of FC2 is slightly higher than the actual value, which can be ascribed to the abnormal voltage drop in the training phase as FC2 worked under more severe operating conditions.

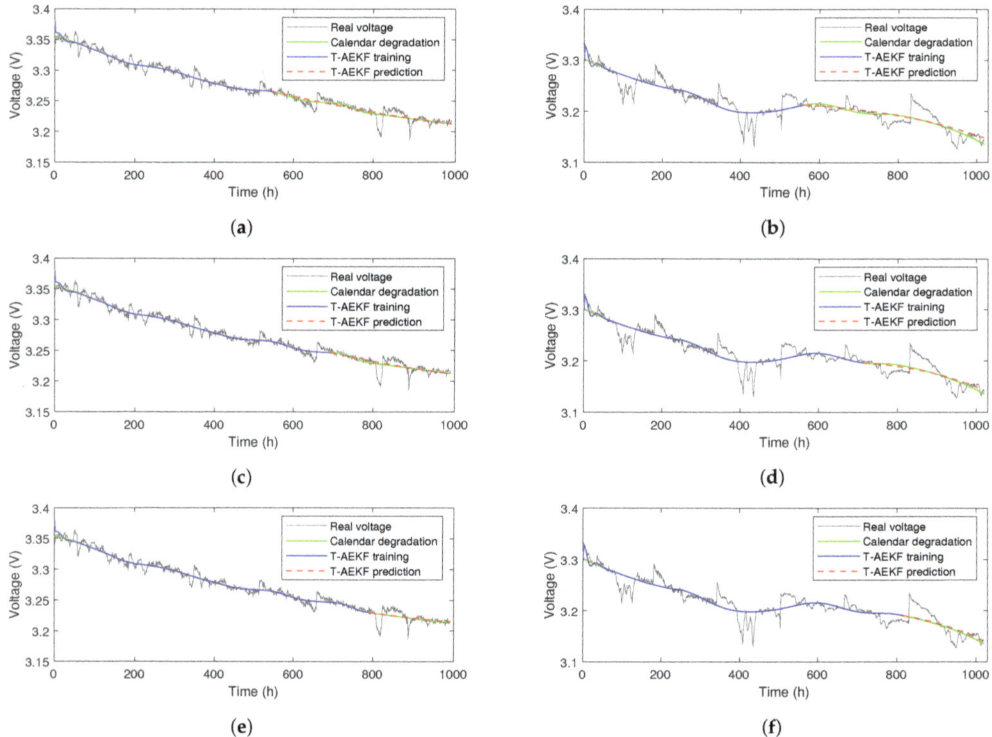

Figure 7. The prediction results based on T-AEKF. (**a**) FC1 with 55% training data; (**b**) FC2 with 55% training data; (**c**) FC1 with 70% training data; (**d**) FC2 with 70% training data; (**e**) FC1 with 80% training data; (**f**) FC2 with 80% training data.

Figure 8a–c demonstrates the estimation results of aging factors for FC1 with 55%, 70%, and 80% training data, and Figure 8d–f demonstrates the estimation results of aging factors for FC2 with 55%, 70%, and 80% training data. The blue lines and the red lines represent the aging factors in the training and the predicting phases, respectively. From Figure 8, we can see that the aging factor α increases slowly as the β decreases with a fluctuation tend. The AEKF algorithm can estimate the aging factor α iteratively so as to update the prediction of the voltage. In the predicting phase, factor γ remains constant. These results show that with the introduction of three-dimensional aging factors (α, β, γ), the proposed calendar aging model can accurately track the overall aging trend both in training and predicting phases.

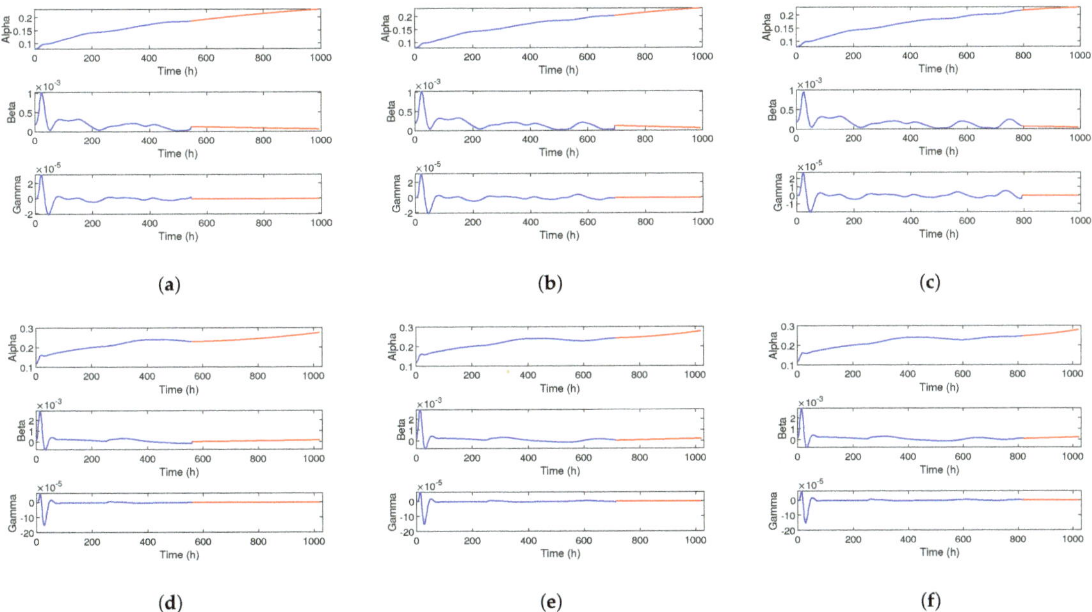

Figure 8. The three-dimensional aging factors of T-AEKF. (**a**) FC1 with 55% training data; (**b**) FC1 with 70% training data; (**c**) FC1 with 80% training data; (**d**) FC2 with 55% training data; (**e**) FC2 with 70% training data; (**f**) FC2 with 80% training data.

Crucially, compared with the real voltage, we note that the spikes and fluctuations exist at the characterization time point for FC1 and FC2 during the aging test, which is regarded as the voltage recovery phenomenon. That is why we built the reversible aging model to capture detailed information for voltage degradation.

3.3. Reversible Aging Voltage Prediction

We deployed the reversible aging model to capture detailed information on the voltage recovery phenomenon. After the voltage decomposition, the sequence of reversible aging voltage is fed into the LSTM network, where the output is the reversible voltage at the next time step. Inspired by [25], we implemented the sliding-window strategy to rebuild the data structure and to improve the prediction accuracy. Since the interruption time of characterization tests is known in advance, we input this information into LSTM as one of the features to make a better prediction, as shown in Figure 6c,d. Additionally, for FC2, as the sharp voltage drop in the two blue dashed boxes is not caused by the normal aging process, we smoothed this abnormal data to improve the prediction performance.

A total of 55% and 80% of the data were used for training and the rest of the data was used for testing. The loss function is the RMSE; the optimizer is Adam. The hyperparameters, including epochs, the number of neurons, and the sliding window size, were initialized arbitrarily, ranging from 50 to 400, 50 to 300, and 10 to 300 with different intervals, respectively. Then, hyperparameters were optimized by the AutoML algorithm with the iteration of 5, automatically. The predicted results of the reversible aging voltage superimposed with the calendar aging voltage will be provided in our final prediction, below.

3.4. Final Aging Voltage Prediction

We added the calendar aging component and the reversible aging component to obtain our final aging voltage prediction. The iterative structure is adopted to realize long-term degradation prediction [21,30]. The predicted values are used as part of the inputs that are fed into the model for forecasting the next step. To verify the advantages of the proposed T-AEKF-LSTM hybrid method, the traditional AEKF method, the LSTM method, and the improved AEKF method based on three-dimensional aging factors (T-AEKF) were used to make a comparison. For the traditional AEKF method, the initial values were the same as the T-AEKF method introduced in Section 3.2. For the LSTM method, the hidden units, epochs, and the sliding windows size were 50, 200, and 20, respectively, which were obtained by testing the performance of LSTM under different configurations.

The predicted results of FC1 under 55%, 70%,and 80% training sets are shown in Figure 9a,c,e, respectively. The traditional AEKF method can only give a linear voltage trend due to its degradation rate remaining constant in the predicting phase. In addition, we find it likely that a bad prediction results when the final point of the training phase is near the abrupt voltage. The LSTM method can predict local nonlinearity which contributes to capturing the voltage recovery phenomenon. However, its output voltage gradually deviates from the measured voltage as time goes on. The T-AEKF method can predict the overall aging trend of PEMFC more accurately than the traditional AEKF method. This decomposition forecasting strategy can prevent the AEKF model from being affected by short-term disturbance and can make the prediction more robust. In addition, three-dimensional aging factors help to model and fit the aging process more accurately, since this scheme can adjust the degradation rate according to the different time. Based on the T-AEKF method and combined with the reversible aging model, the T-AEKF-LSTM method can further capture the voltage recovery information. It can predict the periodic fluctuation in voltage and give a better prediction performance of the aging process for PEMFC compared with other methods.

The prediction results under the dynamic condition for FC2 with 55%, 70%, and 80% training sets are shown in Figure 9b,d,f. It can be found from Figure 9a,d that the AEKF method is not robust enough, as its predicted voltage deviates significantly from the measured voltage. The LSTM method can predict the reversible aging phenomenon after every characterization but fails to trace the aging trend accurately. However, its short-term degradation prediction is more accurate than AEKF and T-AEKF. The T-AEKF can trace the degradation trend better than AEKF but it is not capable of forecasting the reversible aging process. The proposed T-AEKF-LSTM hybrid method can trace the degradation trend and predict reversible voltage components more accurately. It can be noticed that the prediction voltage of the hybrid method will rise slightly at the end of the aging test, which can be ascribed to the memory of the LSTM network, suggesting the occurrence of voltage recovery phenomenon at that time. Thus, the periodic fluctuation in voltage after every characterization test and the nonlinear variation in voltage can be accurately predicted by our hybrid method.

The root mean square error (RMSE) and mean absolute percentage error (MAPE) are used to evaluate the long-term voltage prediction performance [26]. The prediction error is used to evaluate the RUL estimation results. Those criteria are expressed as follows:

$$\text{RMSE} = \sqrt{\frac{1}{N}\sum_{i=1}^{N}(y_i - \hat{y}_i)^2} \tag{20}$$

$$\text{MAPE} = \frac{1}{N}\sum_{1}^{N}\frac{|y_i - \hat{y}_i|}{|y_i|} \times 100 \tag{21}$$

$$\text{Error} = \text{RUL} - \hat{\text{RUL}} \tag{22}$$

where $\hat{y}_i$ is the predicted voltage, and y_i is the measured voltage. RUL represents the actual RUL of the PEMFC, and $\widehat{RUL}$ represents the estimated RUL.

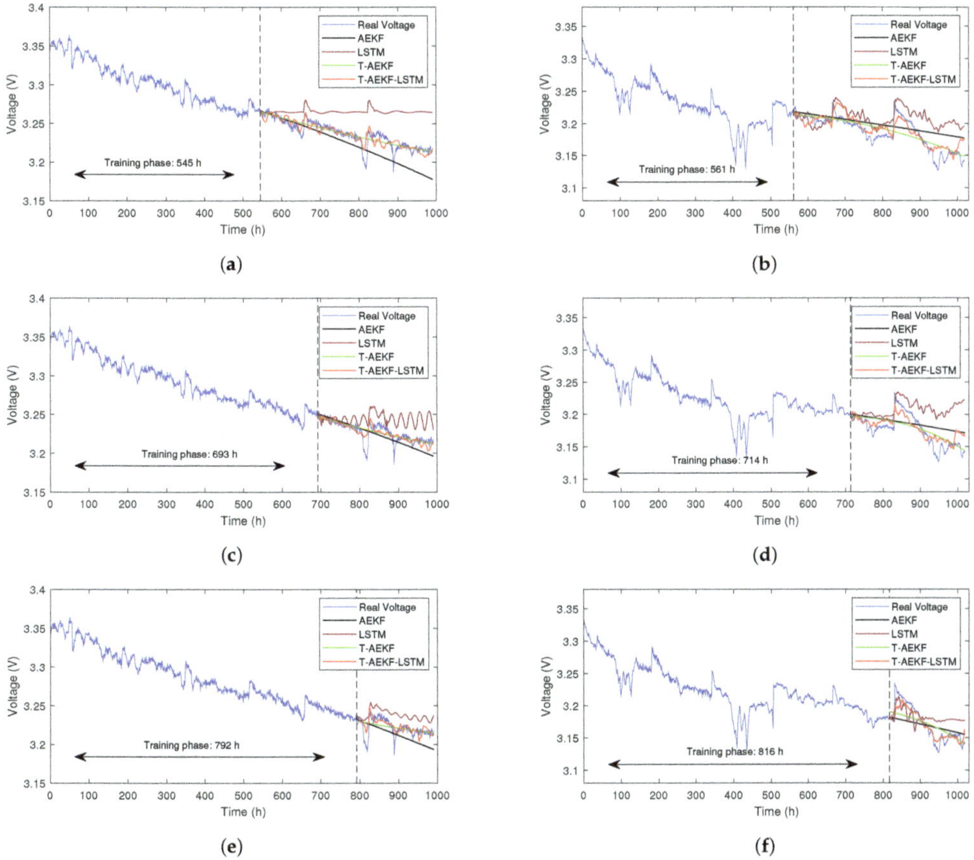

Figure 9. The prediction results of AEKF, LSTM, T-AEKF, and the proposed method (T-AEKF-LSTM). (**a**) FC1 with 55% training data; (**b**) FC2 with 55% training data; (**c**) FC1 with 70% training data; (**d**) FC2 with 70% training data; (**e**) FC1 with 80% training data; (**f**) FC2 with 80% training data.

From the prognostic results for FC1 in Table 3, we can observe that the RMSE and the MAPE of LSTM remain the worst among the four methods. The RMSE and MAPE of T-AEKF are always smaller than AEKF due to the introduction of three-dimensional aging factors as well as the voltage decomposition framework. Since the T-AEKF-LSTM improved the abilities of modeling the reversible aging process based on an LSTM network, it has the lowest prediction error in most cases.

Table 3. The prognostic results for FC1.

	Data	AEKF	LSTM	T-AEKF	T-AEKF-LSTM
	55%	0.0181	0.0338	0.0084	**0.0083**
RMSE	70%	0.0152	0.0232	**0.0087**	0.0092
	80%	0.0151	0.0188	0.0102	**0.0091**
	60%	0.4673	0.9101	0.1994	**0.1913**
MAPE	70%	0.2792	0.5686	**0.1821**	0.2031
	80%	0.4140	0.5214	0.2162	**0.2039**

Following the prognostic results for FC2 shown in Table 4, the T-AEKF-LSTM method has the best performance among the three methods according to its lowest prediction error. Particularly, in FC2, the RMSE and MAPE of T-AEKF-LSTM are much lower than that of T-AEKF, while in FC1, the improvements of T-AEKF-LSTM over the T-AEKF method is not very obvious. A possible explanation may be that the voltage recovery phenomenon of FC2 is more severe than FC1 and it greatly reduces the prediction accuracy of EKF-series-based approaches. In contrast, the reversible aging model can capture this detailed information and can significantly improve the prediction performance. The dramatic fluctuation in voltage can also contribute to the training of the LSTM network. The results above show that the proposed hybrid prognostics method can give a more robust and accurate prediction compared with single AEKF or LSTM methods.

Table 4. The prognostic results for FC2.

	Data	AEKF	LSTM	T-AEKF	T-AEKF-LSTM
	55%	0.0201	0.0295	0.0161	**0.0113**
RMSE	70%	0.0211	0.0340	0.0169	**0.0126**
	80%	0.0221	0.0314	0.0194	**0.0107**
	60%	0.5149	0.7690	0.4104	**0.3027**
MAPE	70%	0.5716	0.8293	0.4244	**0.3251**
	80%	0.5446	0.8506	0.5226	**0.2640**

3.5. RUL Estimation

In this paper, the prediction results of a 55% training set were used to calculate the RUL of PEMFC. The degradation degree 4.0% of the initial voltage was selected as the end of life for fuel cell [33,39]. Since FC2 degrades faster than FC1, 5.0% of the initial voltage was also used to further evaluate the RUL estimation for FC2.

The RUL prediction results based on AEKF, LSTM, T-AEKF, and T-AEKF-LSTM are demonstrated in Table 5. The positive and negative values of the prediction error represent an early prediction or a late prediction, respectively. In order to predict faults in advance, an early prediction is preferred. From Table 5, the RUL estimation error of the proposed T-AEKF-LSTM method is within 30 h and always lower than that of other methods, which indicates that it can give a more accurate RUL estimation among them. The reason for the missing data is that the prediction performances of those methods are too bad to give the prediction errors.

Table 5. The RUL prediction results for FC1 and FC2 (55% training data).

Stack	Degradation Degrees	Actual RUL	AEKF		LSTM		T-AEKF		PAM-ARMA-TDNN [33]		T-AEKF-LSTM (Ours)	
			RUL	Error	RUL	Error	RUL	Error	RUL	Error	RUL	Error
FC1	4.0%	247 h	216 h	31 h	>446 h	-	283 h	−36 h	252 h	−5 h	**244 h**	**3 h**
FC2	4.0%	55 h	204 h	−149 h	207 h	−152 h	172 h	−117 h	156 h	−101 h	**29 h**	**26 h**
	5.0%	359 h	>459 h	-	>459 h	-	386 h	−27 h	381 h	−22 h	**348 h**	**11 h**

As demonstrated in Table 5, the RUL estimation error of the proposed method is always lower than that of the PAM-ARMA-TDNN method [33] for each degradation degree, which verified the advantages of the proposed method. The results above demonstrate the effectiveness and robustness of the proposed method under static and dynamic operating conditions.

4. Conclusions

A robust hybrid prognostic method for PEMFC was proposed in this paper. Considering the voltage recovery phenomenon, a decomposition forecasting framework was established to predict the long-term voltage degradation for PEMFC. Firstly, the original voltage data was decomposed into the calendar aging component and the reversible aging component based on LOESS. Then, we used the AEKF algorithm to predict the overall aging trend of PEMFC based on the calendar aging component. Meanwhile, we introduced three-dimensional aging factors to the physical aging model to better forecast the degradation trend. Next, the LSTM neural network was applied to capture the voltage recovery information through the reversible aging component. Particularly, the AutoML approach based on the genetic algorithm was adopted in the training phase of LSTM for the automatic hyperparameters tuning. The iterative structure was utilized to realize long-term degradation forecasting. The final prediction of the aging voltage can be obtained by combining the two predicted components and we can further realize RUL estimation. We verified the capability of the proposed hybrid prognostic method by two aging datasets under different operating conditions. Experiment results show that the proposed decomposition forecasting framework can combine the advantages of the model-based method for predicting long-term degradation trends and the data-based method for nonlinear modeling ability. In addition, this hybrid method can realize more accurate long-term degradation prediction for PEMFC compared with the single AEKF method or LSTM method. Developing online prognostic methods for PEMFC under high dynamic operating conditions, for example, in automotive applications, is still the major challenge for the prognostic research and it needs further exploration.

Author Contributions: Conceptualization, Z.X. and Y.W.; methodology, Z.X., Y.W. and Y.L.; software, Z.X.; validation, Z.X. and Y.W.; formal analysis, Z.X.; investigation, Z.X. and G.T.; resources, L.M. and G.T.; data curation, L.M. and Y.L.; writing—original draft preparation, Z.X.; writing—review and editing, Z.X. and Y.Z.; visualization, J.T.; supervision, L.M. and G.T.; project administration, L.M. and G.T.; funding acquisition, L.M. and J.T. All authors have read and agreed to the published version of the manuscript.

Funding: This work is supported in part by the Open Research Project of the State Key Laboratory of Industrial Control Technology, Zhejiang University, China (No. ICT2022B31), and in part by the National Key R&D plan of China under Grant 2018YFB1702203.

Institutional Review Board Statement: Not applicable.

Informed Consent Statement: Not applicable.

Data Availability Statement: Not applicable.

Acknowledgments: The authors gratefully acknowledge the support of the National Key R&D plan of China and the Open Research Project of the State Key Laboratory of Industrial Control Technology, Zhejiang University, China.

Conflicts of Interest: The authors declare no conflict of interest.

Abbreviations

The following abbreviations are used in this manuscript:

AEKF	Adaptive extended Kalman filter
AIC	Akaike information criterion
AutoML	Automatic machine learning
BOP	Balance of plant
ECSA	Electrochemical surface area
EIS	Electrochemical impedance spectroscopy
FC	Fuel cell
GDL	Gas diffusion layer
LOESS	Locally weighted regression
LSTM	Long short-term memory (neural network)

MAPE	Mean absolute percentage error
PEMFC	Proton-exchange-membrane fuel cell
RMSE	Root mean-square error
RNN	Recurrent neural network
RUL	Remaining useful life

References

1. Zhang, D.; Li, X.; Wang, W.; Zhao, Z. Internal Characterization-Based Prognostics for Micro-Direct-Methanol Fuel Cells under Dynamic Operating Conditions. *Sensors* **2022**, *22*, 4217. [CrossRef]
2. Liu, H.; Chen, J.; Hissel, D.; Lu, J.; Hou, M.; Shao, Z. Prognostics methods and degradation indexes of proton exchange membrane fuel cells: A review. *Renew. Sustain. Energy Rev.* **2020**, *123*, 109721. [CrossRef]
3. Jouin, M.; Gouriveau, R.; Hissel, D.; Péra, M.C.; Zerhouni, N. Degradations analysis and aging modeling for health assessment and prognostics of PEMFC. *Reliab. Eng. Syst. Saf.* **2016**, *148*, 78–95. [CrossRef]
4. Wu, J.; Yuan, X.Z.; Martin, J.J.; Wang, H.; Zhang, J.; Shen, J.; Wu, S.; Merida, W. A review of PEM fuel cell durability: Degradation mechanisms and mitigation strategies. *J. Power Sources* **2008**, *184*, 104–119. [CrossRef]
5. Yan, X.; Zhou, X.; Zhao, T.; Jiang, H.; Zeng, L. A highly selective proton exchange membrane with highly ordered, vertically aligned, and subnanosized 1D channels for redox flow batteries. *J. Power Sources* **2018**, *406*, 35–41. [CrossRef]
6. Aziz, M.A.; Shanmugam, S. Sulfonated graphene oxide-decorated block copolymer as a proton-exchange membrane: Improving the ion selectivity for all-vanadium redox flow batteries. *J. Mater. Chem. A* **2018**, *6*, 17740–17750. [CrossRef]
7. Rajput, A.; Sharma, P.P.; Yadav, V.; Kulshrestha, V. Highly stable graphene oxide composite proton exchange membrane for electro-chemical energy application. *Int. J. Hydrogen Energy* **2020**, *45*, 16976–16983. [CrossRef]
8. Chen, K.; Laghrouche, S.; Djerdir, A. Fuel cell health prognosis using Unscented Kalman Filter: Postal fuel cell electric vehicles case study. *Int. J. Hydrogen Energy* **2019**, *44*, 1930–1939. [CrossRef]
9. Cigolotti, V.; Genovese, M.; Fragiacomo, P. Comprehensive review on fuel cell technology for stationary applications as sustainable and efficient poly-generation energy systems. *Energies* **2021**, *14*, 4963. [CrossRef]
10. Zhang, X.; Yang, D.; Luo, M.; Dong, Z. Load profile based empirical model for the lifetime prediction of an automotive PEM fuel cell. *Int. J. Hydrogen Energy* **2017**, *42*, 11868–11878. [CrossRef]
11. Peng, C.; Chen, Y.; Chen, Q.; Tang, Z.; Li, L.; Gui, W. A Remaining Useful Life prognosis of turbofan engine using temporal and spatial feature fusion. *Sensors* **2021**, *21*, 418. [CrossRef]
12. Yang, J.; Peng, Y.; Xie, J.; Wang, P. Remaining Useful Life Prediction Method for Bearings Based on LSTM with Uncertainty Quantification. *Sensors* **2022**, *22*, 4549. [CrossRef] [PubMed]
13. Bressel, M.; Hilairet, M.; Hissel, D.; Bouamama, B.O. Extended Kalman filter for prognostic of proton exchange membrane fuel cell. *Appl. Energy* **2016**, *164*, 220–227. [CrossRef]
14. Sutharssan, T.; Montalvao, D.; Chen, Y.K.; Wang, W.C.; Pisac, C.; Elemara, H. A review on prognostics and health monitoring of proton exchange membrane fuel cell. *Renew. Sustain. Energy Rev.* **2017**, *75*, 440–450. [CrossRef]
15. Zhang, X.; Pisu, P. An Unscented Kalman Filter Based Approach for the Health Monitoring and Prognostics of a Polymer Electrolyte Membrane Fuel Cell. In Proceedings of the Annual Conference of the PHM Society, Minneapolis, MN, USA, 23–27 September 2012.
16. Dhanushkodi, S.; Schwager, M.; Todd, D.; Merida, W. PEMFC durability: Spatially resolved Pt dissolution in a single cell. *J. Electrochem. Soc.* **2014**, *161*, F1315. [CrossRef]
17. Bi, W.; Fuller, T.F. Modeling of PEM fuel cell Pt/C catalyst degradation. *J. Power Sources* **2008**, *178*, 188–196. [CrossRef]
18. Polverino, P.; Pianese, C. Model-based prognostic algorithm for online RUL estimation of PEMFCs. In Proceedings of the 2016 3rd Conference on Control and Fault-Tolerant Systems (SysTol), Barcelona, Spain, 7–9 September 2016; IEEE: Piscataway Township, NJ, USA, 2016; pp. 599–604.
19. Jouin, M.; Gouriveau, R.; Hissel, D.; Péra, M.C.; Zerhouni, N. Prognostics of PEM fuel cell in a particle filtering framework. *Int. J. Hydrogen Energy* **2014**, *39*, 481–494. [CrossRef]
20. Li, B.; Bei, S. Estimation algorithm research for lithium battery SOC in electric vehicles based on adaptive unscented Kalman filter. *Neural Comput. Appl.* **2019**, *31*, 8171–8183. [CrossRef]
21. Silva, R.; Gouriveau, R.; Jemei, S.; Hissel, D.; Boulon, L.; Agbossou, K.; Steiner, N.Y. Proton exchange membrane fuel cell degradation prediction based on adaptive neuro-fuzzy inference systems. *Int. J. Hydrogen Energy* **2014**, *39*, 11128–11144. [CrossRef]
22. Liu, H.; Chen, J.; Hissel, D.; Su, H. Short-term prognostics of PEM fuel cells: A comparative and improvement study. *IEEE Trans. Ind. Electron.* **2018**, *66*, 6077–6086. [CrossRef]
23. Morando, S.; Jemei, S.; Gouriveau, R.; Zerhouni, N.; Hissel, D. Fuel cells prognostics using echo state network. In Proceedings of the IECON 2013-39th Annual Conference of the IEEE Industrial Electronics Society, Vienna, Austria, 10–13 November 2013; IEEE: Piscataway Township, NJ, USA, 2013; pp. 1632–1637.
24. Hua, Z.; Zheng, Z.; Pahon, E.; Péra, M.C.; Gao, F. Remaining useful life prediction of PEMFC systems under dynamic operating conditions. *Energy Convers. Manag.* **2021**, *231*, 113825. [CrossRef]

25. Ma, R.; Yang, T.; Breaz, E.; Li, Z.; Briois, P.; Gao, F. Data-driven proton exchange membrane fuel cell degradation predication through deep learning method. *Appl. Energy* **2018**, *231*, 102–115. [CrossRef]

26. Liu, H.; Chen, J.; Hissel, D.; Su, H. Remaining useful life estimation for proton exchange membrane fuel cells using a hybrid method. *Appl. Energy* **2019**, *237*, 910–919. [CrossRef]

27. Li, Z.; Zheng, Z.; Outbib, R. Adaptive prognostic of fuel cells by implementing ensemble echo state networks in time-varying model space. *IEEE Trans. Ind. Electron.* **2019**, *67*, 379–389. [CrossRef]

28. Ma, R.; Xie, R.; Xu, L.; Huangfu, Y.; Li, Y. A Hybrid Prognostic Method for PEMFC with Aging Parameter Prediction. *IEEE Trans. Transp. Electrif.* **2021**, *7*, 2318–2331. [CrossRef]

29. Jouin, M.; Gouriveau, R.; Hissel, D.; Péra, M.C.; Zerhouni, N. Joint particle filters prognostics for proton exchange membrane fuel cell power prediction at constant current solicitation. *IEEE Trans. Reliab.* **2015**, *65*, 336–349. [CrossRef]

30. Morando, S.; Jemei, S.; Hissel, D.; Gouriveau, R.; Zerhouni, N. Proton exchange membrane fuel cell ageing forecasting algorithm based on Echo State Network. *Int. J. Hydrogen Energy* **2017**, *42*, 1472–1480. [CrossRef]

31. Kimotho, J.K.; Meyer, T.; Sextro, W. PEM fuel cell prognostics using particle filter with model parameter adaptation. In Proceedings of the 2014 International Conference on Prognostics and Health Management, Zhangjiajie, China 24–27 August 2014; IEEE: Piscataway Township, NJ, USA, 2014, pp. 1–6.

32. Deng, L.; Shen, W.; Wang, H.; Wang, S. A rest-time-based prognostic model for remaining useful life prediction of lithium-ion battery. *Neural Comput. Appl.* **2021**, *33*, 2035–2046. [CrossRef]

33. Zhou, D.; Al-Durra, A.; Zhang, K.; Ravey, A.; Gao, F. Online remaining useful lifetime prediction of proton exchange membrane fuel cells using a novel robust methodology. *J. Power Sources* **2018**, *399*, 314–328. [CrossRef]

34. Mao, L.; Jackson, L. IEEE 2014 Data Challenge Data. 2016. Available online: https://repository.lboro.ac.uk/articles/dataset/IEEE_2014_Data_Challenge_Data/3518141/1 (accessed on 20 October 2021).

35. Liu, J.; Li, Q.; Chen, W.; Yan, Y.; Qiu, Y.; Cao, T. Remaining useful life prediction of PEMFC based on long short-term memory recurrent neural networks. *Int. J. Hydrogen Energy* **2019**, *44*, 5470–5480. [CrossRef]

36. Taieb, S.B.; Bontempi, G.; Atiya, A.F.; Sorjamaa, A. A review and comparison of strategies for multi-step ahead time series forecasting based on the NN5 forecasting competition. *Expert Syst. Appl.* **2012**, *39*, 7067–7083. [CrossRef]

37. Bressel, M.; Hilairet, M.; Hissel, D.; Bouamama, B.O. Remaining useful life prediction and uncertainty quantification of proton exchange membrane fuel cell under variable load. *IEEE Trans. Ind. Electron.* **2016**, *63*, 2569–2577. [CrossRef]

38. Fowler, M.W.; Mann, R.F.; Amphlett, J.C.; Peppley, B.A.; Roberge, P.R. Incorporation of voltage degradation into a generalised steady state electrochemical model for a PEM fuel cell. *J. Power Sources* **2002**, *106*, 274–283. [CrossRef]

39. Pan, R.; Yang, D.; Wang, Y.; Chen, Z. Performance degradation prediction of proton exchange membrane fuel cell using a hybrid prognostic approach. *Int. J. Hydrogen Energy* **2020**, *45*, 30994–31008. [CrossRef]

40. Burnham, K.P.; Anderson, D.R. Multimodel inference: Understanding AIC and BIC in model selection. *Sociol. Methods Res.* **2004**, *33*, 261–304. [CrossRef]

41. Bodenhofer, U. Genetic Algorithms: Theory and Applications. Lecture notes, Fuzzy Logic Laboratorium Linz-Hagenberg, Winter, 2004. 2003. Available online: http://www.flll.jku.at/div/teaching/Ga/notes.pdf (accessed on 20 October 2021).

42. Wang, Y.; Zhang, C.; Chen, Z. An adaptive remaining energy prediction approach for lithium-ion batteries in electric vehicles. *J. Power Sources* **2016**, *305*, 80–88. [CrossRef]

43. Sundermeyer, M.; Schlüter, R.; Ney, H. LSTM neural networks for language modeling. In Proceedings of the Thirteenth Annual Conference of the International Speech Communication Association, Portland, OR, USA, 9–13 September 2012.

44. Siami-Namini, S.; Tavakoli, N.; Namin, A.S. The performance of LSTM and BiLSTM in forecasting time series. In Proceedings of the 2019 IEEE International Conference on Big Data (Big Data), Los Angeles, CA, USA, 9–12 December 2019; IEEE: Piscataway Township, NJ, USA, 2019; pp. 3285–3292.

Article

An Efficient and Intelligent Detection Method for Fabric Defects based on Improved YOLOv5

Guijuan Lin [1,*], Keyu Liu [1], Xuke Xia [2] and Ruopeng Yan [1]

[1] School of Mechanical and Automotive Engineering, Xiamen University of Technology, Xiamen 361024, China
[2] Quanzhou Institute of Equipment Manufacturing, Haixi Institute, Chinese Academy of Sciences, Jinjiang 362216, China
* Correspondence: forestlgj@126.com

Abstract: Limited by computing resources of embedded devices, there are problems in the field of fabric defect detection, including small defect size, extremely unbalanced aspect ratio of defect size, and slow detection speed. To address these problems, a sliding window multihead self-attention mechanism is proposed for the detection of small targets, and the Swin Transformer module is introduced to replace the main module in the original YOLOv5 algorithm. First, to reduce the distance between several scales, the weighted bidirectional feature network is employed on embedded devices. In addition, it is helpful to improve the perception of small-target faults by incorporating a detection layer to achieve four-scale detection. At last, to improve the learning of positive sample instances and lower the missed detection rate, the generalized focal loss function is finally implemented on YOLOv5. Experimental results show that the accuracy of the improved algorithm on the fabric dataset reaches 85.6%, and the mAP is increased by 4.2% to 76.5%, which meets the requirements for real-time detection on embedded devices.

Keywords: deep learning; computer vision; fabric detection; Swin Transformer; YOLOv5

Citation: Lin, G.; Liu, K.; Xia, X.; Yan, R. An Efficient and Intelligent Detection Method for Fabric Defects based on Improved YOLOv5. *Sensors* **2023**, *23*, 97. https://doi.org/10.3390/s23010097

Academic Editors: Yong Liu and Xingxing Zuo

Received: 29 November 2022
Revised: 15 December 2022
Accepted: 19 December 2022
Published: 22 December 2022

1. Introduction

Fabric defect detection is a core and important link in the entire textile quality production process. At present, the detection effect of existing methods for some small and widely distributed defects cannot meet the requirements of manufacturers [1]. Thus, it is of great importance to use accurate and efficient detection methods to improve the detection and identification of fabric defects.

In the field of fabric defect detection, traditional image processing methods are only suitable for detecting solid-color fabrics [2–6]. For fabrics with complex texture patterns, such as printed and jacquard fabrics, defect types are difficult to distinguish, especially for the detection of small defect targets. Traditional visual processing methods have been difficult to meet the needs of enterprises. Combining traditional vision techniques with deep learning for appearance defect detection of various objects has achieved considerable results. Reference [7] applies supervised learning to fabric defect detection by extracting effective features. Reference [8] trains a stacked denoising autoencoder based on Fisher's criterion by using defective and nondefective samples, and uses the residual threshold to locate defects. With the development of the multiscale detection network, the detection methods of fabric multidefects have been proposed one after another. Zhang et al. [9] proposed a method for automatic location of fabric defects based on YOLO, which can meet the classification and detection of colored fabrics. Wang [10] proposed a detection algorithm based on the DeeplabV3+ model, which used the advantages of multiscale target detection and improved detection accuracy while reducing the network model parameters, as well as the ability to detect small-sized targets. Good results are obtained in the defect dataset.

In practical production applications, there is the problem of data imbalance in the defects of fabrics. For example, the number of fabric defect samples is small, some span of fabric defect sizes is large, and the number of small defects is large. Reference [11] divides and synthesizes multiple reconstruction residual images to obtain new defect detection results. This method can reduce the difficulty of collecting defect samples. Huang et al. [12] first take untrained fabric data as input. The output of the segmentation model is then used as raw material for the decision model. This approach requires only a few defect samples and can train a more accurate detection model.

In this study, we propose an efficient and intelligent detection method for fabric defects. The main contributions of this paper are as follows:

1. Based on the Transformer structure, we optimize the YOLOv5 v6.1 algorithm with the Swin Transformer as the backbone, and the introduction of a multiwindow sliding self-attention mechanism complements the convolutional network to improve classification accuracy.
2. In the neck layer, the BiFPN is used to replace the original FPN to enhance the fusion of semantic information between different layers, and a small-target detection layer is added to improve the detection effect of the model on small targets.
3. We introduce the generalized focal loss function to enhance the model's instance learning of positive samples, in order to alleviate the problems caused by the imbalance of fabric samples.
4. Finally, we conducted ablation experiments and an in-depth analysis of the impact of the above-mentioned improved methods and several attention mechanisms on detection accuracy and real-time performance. Our proposed method outperforms current popular object detection models on a self-created fabric dataset.

In the remainder of this paper, the second section summarizes the YOLOv5 algorithm and the optimization method of this paper. Section 3 presents the fabric dataset, experimental details, and concrete results. The last section is the conclusion of this paper and the prospect of follow-up work.

2. Materials and Methods

There are generally two target detection methods at present: two stage and one stage [13]. Two stage first generates a series of sample candidate boxes through the algorithm, and then performs classification through a series of convolution operations. Mainstream two-stage algorithms include the R-CNN network proposed by Girshick [14] in 2014. The two-stage network is characterized by high accuracy of positioning and detection, but due to the complex network structure and poor real-time performance, it is not effective for rapid detection in the industry. One-stage methods do not need to select the sample candidate frame. They can directly obtain the coordinates and type of the target, which not only has better real-time performance, but also has advantages in small-target detection [15–18].

2.1. Structure of the YOLOv5

You only look once [19] (YOLO) is a single-stage target detection algorithm based on full convolution. YOLO [20] can predict the entire picture and give all the detection results at one time. The YOLOv5 algorithm is the fifth version of the YOLO algorithm launched by the Ultralytics LLC team. Its model has the characteristics of simplicity, speed, and portability.

The YOLOv5 algorithm framework consists of the input, the backbone, the neck, and the prediction. The structure diagram is shown in Figure 1.

The input of YOLOv5 mainly uses two sections: mosaic data enhancement and adaptive anchor frame. Mosaic data enhancement can read four pictures at a time; randomly scale, crop, and arrange each picture; and finally, randomly splice them together. This can greatly enrich the number of datasets and make the entire network more robust. For pictures of different sizes, first, the size of the input image is adjusted to a uniform set size, and in the process of scaling and filling, the original samples are adaptively populated and then sent to the backbone network.

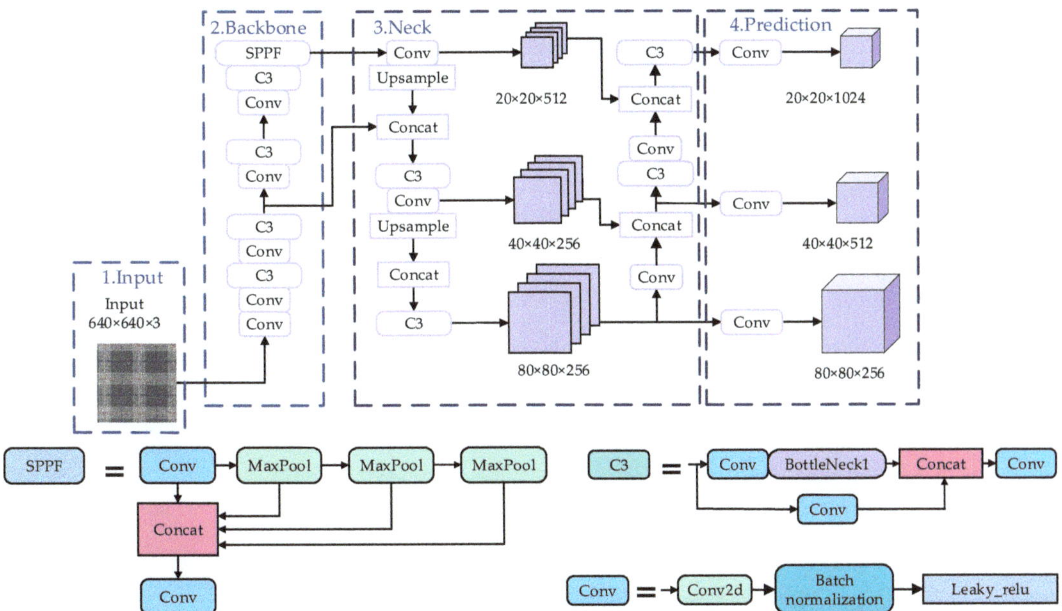

Figure 1. Network structure for YOLOv5.

The backbone network consists of a Conv structure and C3 structure. The CSP [21] structure solves the problem of excessive calculation in reasoning from the perspective of model structure design. The role of the SPPF layer increases the receptive field through convolution operations and maximum pooling. This module can strengthen the nonlinear expression capability.

The feature fusion layer neck adopts the structure of PAN + FPN [22], which fuses different detection layers with the main feature layer. The FPN only enhances the transmission of semantic information, but the ability to transmit low-level positioning information is not strong. On the basis of the FPN, PAN adds a bottom-up pyramid through a 3×3 convolution to enhance the transmission of positioning information. On the prediction side, this module uses GIoU [23] to calculate the loss value of the bounding box [24].

2.2. Swin Transformer Model

The Swin Transformer [25] model consists of multilayer perceptron (MLP), layer normalization (LM), window multihead self-attention (W-MSA), and sliding window multihead self-attention(SW-MSA). The structure of the Swin Transformer is similar to the traditional residual structure, so it can be directly used in convolutional networks. The Swin Transformer's structure in the backbone layer is shown in Figure 2.

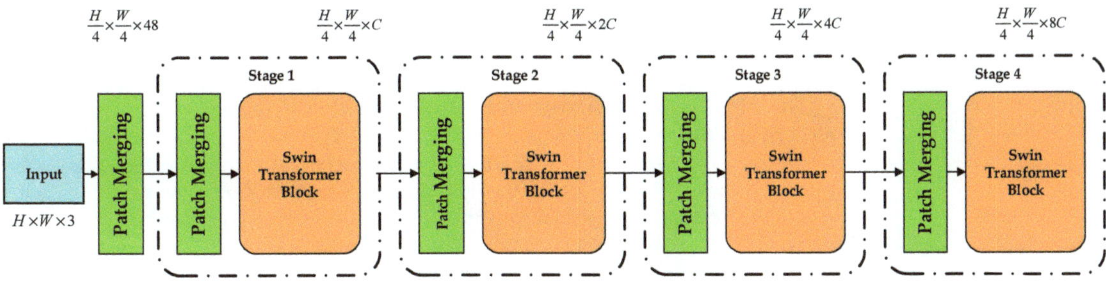

Figure 2. Swin Transformer network structure.

The workflow is as follows: (1) The Swin Transformer first inputs an RGB three-channel image of size H × W. It divides the input image into 4 × 4 patches through the segmentation layer, and its feature dimension is divided into h/4 × h/4. (2) In the process of Stage 1, the dimension of the output is changed to C through linear embedding, and then dispatches the Transformer block for Stage 2. (3) The purpose of Stage 2 to Stage 4 is the same. After the 2 × 2 adjacent blocks are spliced through image block patch merging, the spliced high-dimensional features are then reduced in dimension through a convolution, and the output dimension becomes 2C. Then, Stage 3 is repeated 6 times, and the output after dimensionality reduction is 4C. Stage 4 is repeated twice, and the output dimension becomes 8C.

The standard Transformer encoder consists of a multihead self-attention mechanism and a multilayer perceptron. It uses the layer norm at the beginning of the module, and then uses residual connections between each module. Figure 3 shows its structure. For the Swin Transformer backbone, the layer and calculation formula is:

$$\hat{z}^l = W - MSA\,(LN\,(z^{l-1})) + z^{l-1} \tag{1}$$

$$z^l = MLP(LN(\hat{z}^l)) + \hat{z}^l \tag{2}$$

$$\hat{z}^{l+1} = SW - MSA\,(LN\,(z^l)) + z^l \tag{3}$$

$$z^{l+1} = MLP(LN(\hat{z}^{l+1})) + \hat{z}^{l+1} \tag{4}$$

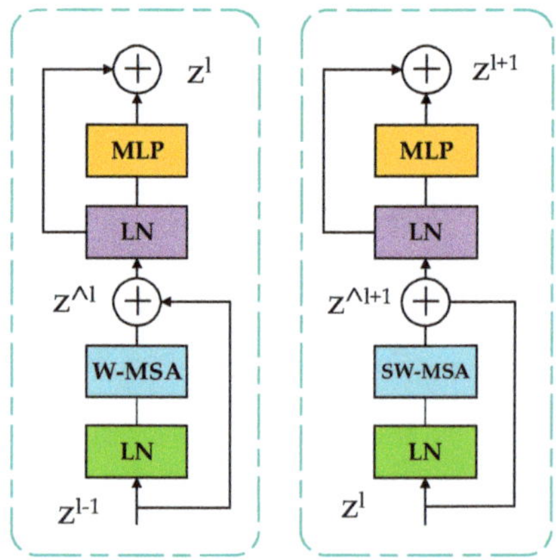

Figure 3. Swin Transformer's backbone network.

In Formulas (1) – (4) above, $\hat{z}$ represents the feature output by MLP, and z^l is the output feature of W-MAS. W-SMA is a traditional window-partitioned self-attention mechanism, while SW-MSA represents a multihead self-attention mechanism using shifted window partitions. By introducing adjacent nonoverlapping windows in the upper layer, the connection between each layer is increased and the classification accuracy is improved.

2.3. Multiwindow Sliding Self-Attention Mechanism

The self-attention in Transformer is the key module of the algorithm. The traditional Transformer structure uses a global self-attention mechanism [26], which greatly increases the amount of computation. The self-attention mechanism based on the local window proposed by the Swin Transformer can make the computing window evenly divide the image in a nonoverlapping manner, and the computational complexity of the window based on H × W image blocks is much smaller than that of the global attention mechanism.

SW-MSA is not limited to different windows for information exchange. As shown in Figure 4, in the first layer, the normal window division method, but at the l + 1 layer, the door division is moved, and the use of the mobile mouthpiece splitting method makes it possible to connect with each other without overlapping the entrance, greatly increasing the reception area of the mouth and increasing the special delivery ability.

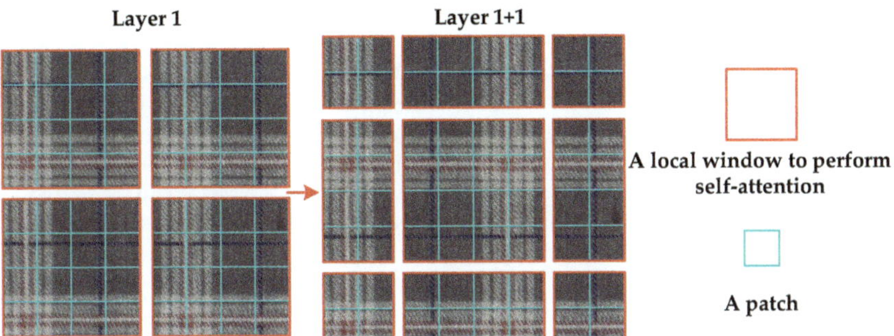

Figure 4. SW-MSA window moves.

2.4. Multiscale Feature Fusion Feature Pyramid Network

In a convolutional neural network, the feature maps obtained by convolutional layers with different parameters contain the feature information of different targets. The feature map obtained after deep convolution has higher resolution, mainly contains position information, but lacks semantic information, while the content obtained by shallow convolution is just the opposite of the former. Therefore, it is necessary to fuse the feature information of the deep feature map and the shallow feature map. The original YOLOv5 algorithm bidirectionally fuses the FPN and PAN in the neck layer to extract the information from different feature layers.

The size of some defects in the fabric is too small, which will cause the feature information extracted by YOLOv5 to ignore the small defect information. In order to strengthen the feature fusion between different scales and increase the detection accuracy, this paper introduces a weighted bidirectional feature network [27] (bidirectional feature network), which uses weighted feature fusion and cross-scale connections to obtain multiple levels. The global features of semantic information can strengthen the recognition accuracy of small object defects. The structure of the BiFPN is shown in Figure 5.

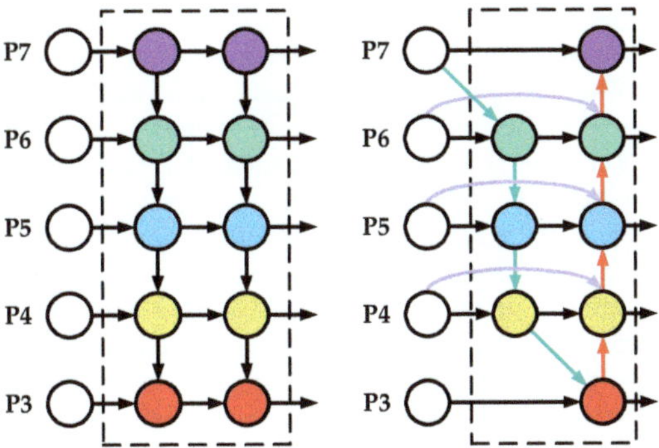

Figure 5. Structure of BiFPN.

The model is used for multidefect detection, and the target defect size on the fabric is not the same. Compared with common target detection algorithms (feature pyramid networks (FPNs)), the BiFPN uses skip connections to lighten the network. An attention mechanism is added to extract deeper feature information. Feature fusion is performed bidirectionally through upsampling and downsampling to improve the feature fusion effect between different layers.

Feature maps of 3 different sizes are used in the original YOLOv5 network results to detect objects with inconsistent sizes. When we set the input image size to 640×640, after a series of convolution and sampling operations, the size of the output detection feature map is 20×20, 40×40, and 80×80, which can detect 32×32, 16×16, and 8×8 targets respectively.

Considering that the pixel size occupied by the small fabric defect targets is extremely small, to further strengthen the detection accuracy of small objects, we improve the algorithm to increase the number of upsamples to improve the lower-level feature information. Four-scale detection is formed by adding a 160×160 detection layer. The improved detection was shown in Figure 6. For the 640×640 input image, the detection layer can detect the 4×4 size target, which further improves the ability to extract small pixel defects in the fabric image.

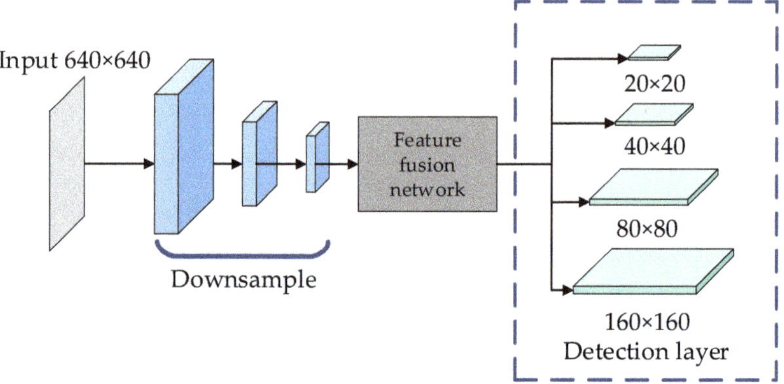

Figure 6. Improved four-scale detection layer.

2.5. Improvement of Loss Function

The loss function is obtained by calculating the error between the real frame and the predicted frame of the positive sample. The loss function of YOLOv5 mainly includes the classification loss function, the localization loss function, and the confidence loss function. In fabric defect detection, the area where the defect is located accounts for a small proportion of the entire fabric image. During training, we regard the area containing the defect as a positive sample and the normal area as a negative sample, which will lead to defective samples. The number of defects is much lower than the normal number of samples in the image area of the fabric. The loss value obtained is mostly the background loss of negative samples. Therefore, focal loss [28] is often cited to balance the number of foreground and background detection samples.

However, there are two problems with focal loss. One is that the focal loss function calculates the positioning quality score (IoU score) and the classification score (classification score) separately during training, but the two are comprehensively multiplied during testing as nonmaximum suppression (nonmaximum suppression (NMS)) sorting basis. This method will lead to a large error between training and testing, which will lead to a decrease in the performance of the detection model and ultimately affect the detection accuracy.

To solve the above problems, the IoU score is merged with the classification score. Since the combined category label becomes a continuous value of 0–1, and focal loss only calculates discrete labels of 0 or 1, this paper introduces generalized focal loss [29] to realize the fused representation of the IoU score and classification score. Its calculation formula is as follows:

$$\text{QFL}(\sigma) = -\,|\,y - \sigma\,|^{\beta}\,((1 - y)\log(1 - \sigma) + y\log(\sigma)) \tag{5}$$

$$\text{GEL}\,(p_{y1}, p_{yr}) = -\,|\,y - (y_1 p_{y1} + y_r p_{yr})\,|^{\beta}\,((y_r - y)\log(p_{y1}) + (y - y_l)\log(p_{yr})) \tag{6}$$

y is the overall label of the detection target, and y_r and y_l are the true values of classification and regression. p_{y1} and p_{yr} are the predicted values corresponding to the former, and β is the hyperparameter. Quality focal loss and distribution focal loss make up this function. Among them, quality focal loss uses hyperparameters to ensure the balance between the number of categories. Its formula is as follows: σ stands for the predicted value, and y stands for the quality label between 0 and 1.

3. Experiments

3.1. Experimental Platform

The experiments in this paper are completed in the environment of Table 1.

Table 1. The experimental environment.

Project	Hardware Specifications (Software Version)
operating system	Ubuntu18.04
CPU	AMD Ryzen5 5600X
GPU	NVIDIA GeForce RTX 3060TI
Software environment	Pytorch 1.7.0, Python3.9, OpenCV 4.6, CUDA 11.6, CuDNN 8.4.0

3.2. Dataset Description

The data set used in the experiment were taken by Alibaba Cloud Tianchi [30] in a textile workshop in Guangdong Province. After manual sorting and selection, 10,321 pictures were selected, including 8 kinds of defects. They are ColorFly, Singeing, Knot, Warp Loosening, ColorOut, Warper's Knot, Hole, and Coarse. An example of each type of fabric defect is shown Figure 7. The specific number of fabric faults is shown in Figure 8.

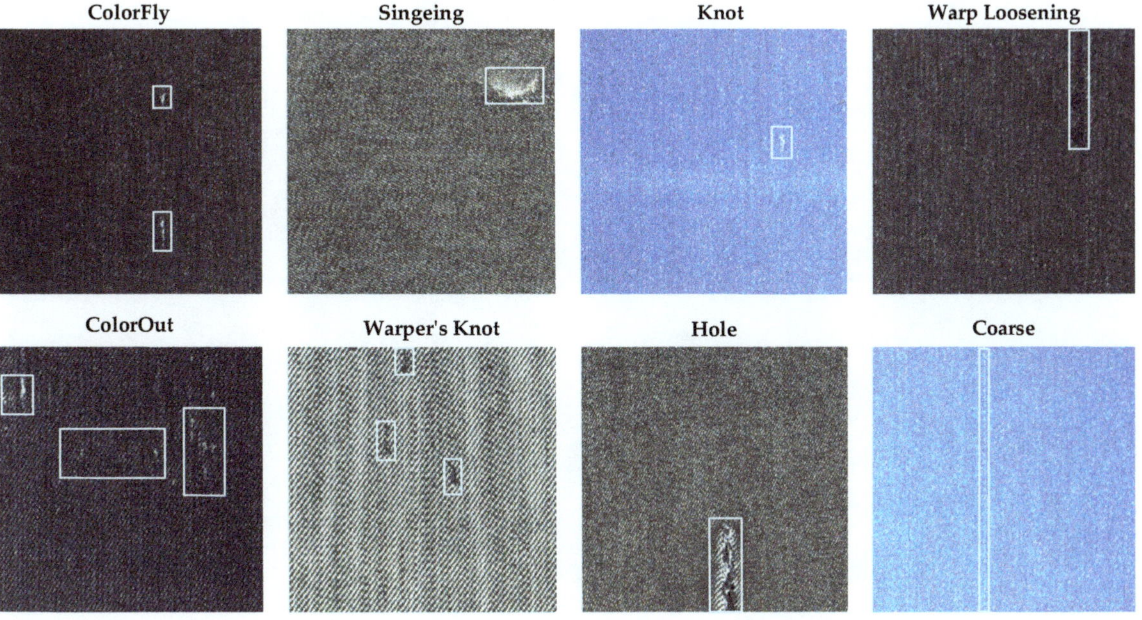

Figure 7. Examples of different types of fabric defects.

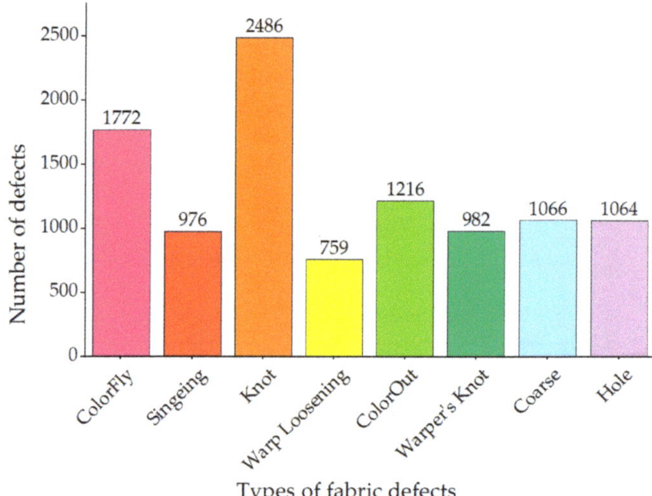

Figure 8. Statistics of fabric defect types.

Figure 9 illustrates the proportion of fabric defects in the entire fabric area. It can be seen that in the fabric data set, most of the defects are small-sized defects, and the features are difficult to capture, and the aspect ratio of the defects is widely distributed, so detection is difficult.

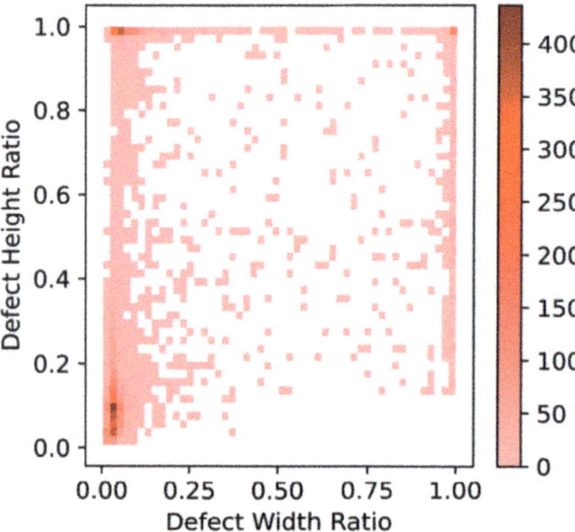

Figure 9. Proportion of defect size.

For the sake of the training accuracy of the model and to prevent overfitting, this paper divides the fabric data set into training, validation, and test set. Its division proportion is 80%, 10%, and 10%.

Due to the low number of certain kinds of defects in the original fabric dataset, it may cause underfitting of this type of defect. In this paper, methods such as flipping, zooming, adding noise, splicing, and mosaic enhancement are used to expand the number of some fabric images. The effect of mosaic data enhancement is shown in Figure 10. Combining four different fabric pictures enables the model to learn various types of features during each training, thereby improving the detection and generalization capabilities of the model, and consecutive effects of the imbalance in the number of fabrics.

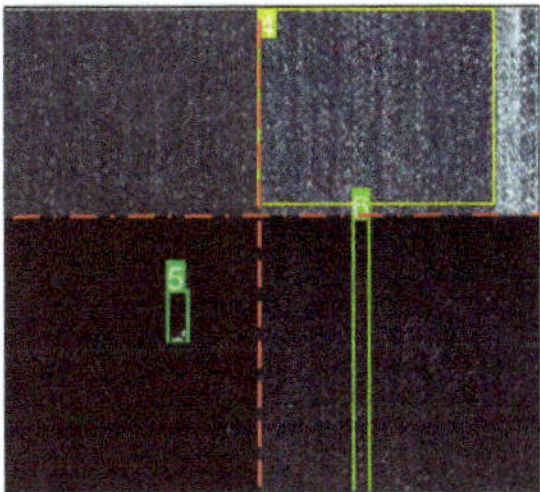

Figure 10. Mosaic data enhancement.

3.3. Evaluation Indicators and Experiment

To demonstrate the effectiveness of the modified algorithm for fabric detection, this experiment adopts precision rate (P), recall rate (R), and mean average precision (mAP) as evaluation indicators of the model.

The mAP is an index to comprehensively measure the accuracy of model detection, and it is the most important index in target detection. Its specific calculation formula is as follows:

$$\text{Precision} = \text{TP}/(\text{TP} + \text{FP}) \tag{7}$$

$$\text{Recall} = \text{TP}/(\text{TP} + \text{FN}) \tag{8}$$

$$\text{mAP} = \sum \text{AP}/\text{C} \tag{9}$$

TP is the number of correctly detected targets, FP is the number of falsely detected targets, FN is the number of missed detections, C is the total number of defect categories, and $\sum$AP is the sum of the precision values of all defect categories.

In addition, this experiment also introduces the frames per second (FPS) and the parameter size of the model as one of the evaluation indicators of the model performance. The higher the speed, the more it can satisfy the needs of real-time detection of fabrics.

In the training phase of this model, the input size of the fabric image is changed to 640×640, the initial learning rate is set to 0.0005, the optimizer selection is SGD, and the batch size is changed to 8.

Figure 11 shows the variation of loss value with epoch during training of the benchmark network and the improved network. Based on the loss function graph, we know that the loss decreases rapidly during the 25 epochs of training, and the loss decreases gradually and becomes smooth after 150 epochs after training. It can be shown that the network training has not occurred overfitting. The loss of the improved model training and validation has a higher drop rate, indicating that the training effect of the improved model is better than the baseline model.

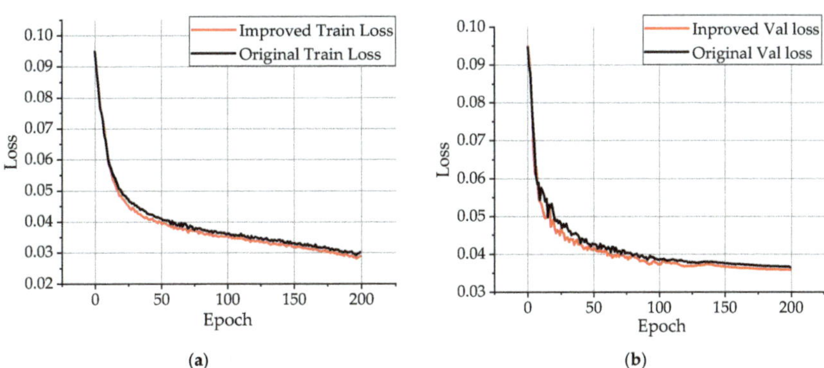

(a)   (b)

Figure 11. Loss function graph. (**a**) Train loss; (**b**) Val loss.

3.4. Loss Function Effect Verification

In order to solve the problem of fabric imbalance, this paper first performs mosaic data enhancement, translation, flip, and other operations on the data to improve the generalization of the network by increasing the number. Secondly, the generalized focal loss function is used to enhance the model's learning of positive samples. In this section, focal loss and generalized focal loss are introduced in the benchmark network to explore the impact of the improvement of the loss function on the detection accuracy.

It can be seen from Table 2 that the generalized focal loss function introduced in this paper can improve the classification effect. The generalized focal loss function increases the learning weight of positive samples and reduces the learning weight of useless negative samples, so as to resist the imbalance caused by positive and negative samples. Secondly, the generalized focal loss function predicts the results through discretization to improve the value of the IoU.

Table 2. Loss function optimization result.

Algorithm	mAP (%)
YOLO	72.3
YOLO + Focal Loss	72.4 (+0.1)
YOLO+ Generalized Focal Loss	72.8 (+0.5)

3.5. Ablation Experiment

Through the ablation experiments on the improved algorithm structure, the effectiveness of our proposed modules and improvements in the performance of fabric defect detection networks is verified. In this paper, YOLOv5 is used as the benchmark network and the structure using Swin Transformer as the backbone and introducing the BiFPN and the small-target detection layer structure called YOLO-SB. On this basis, the network obtained by removing the BiFPN and introducing generalized focal loss is called YOLO-SL. On the basis of the BiFPN and small-target layer, the network that only introduces generalized focal loss is called YOLO-LB. The network introduced by all the improvements is called YOLO-TLB. The experimental results are shown in Table 3. The mAP results of different defect types are shown in Table 4.

Table 3. Ablation experiment.

Algorithm	With Swin T	With Loss	With BiFPN	mAP@0.5	Recall	Weight (MB)
YOLO-LB		√	√	73.8	71.4	14.9
YOLO-SL	√	√		74.6	70.6	18.3
YOLO-SB	√		√	74.8	71.8	20.0
YOLO-TLB	√	√	√	75.9	73.1	20.1

Table 4. Results for each fabric defect category.

Defect Type	mAP@0.5				
	YOLOv5	YOLO-LB	YOLO-SL	YOLO-SB	YOLO-TBL
ColorFly	77.6	79.8	76.9	79.1	81.9
Singeing	60.3	68.8	67.5	64.9	66.7
Knot	72.7	68.2	65.8	73.1	73.7
Warp Loosening	58.5	61.7	66.8	66.8	62.6
ColorOut	88.0	89.1	91.2	89.6	91.3
Warper's Knot	53.6	53.2	59.0	55.4	55.2
Hole	73.7	79.5	78.2	76.9	82.1
Coarse	92.7	90.2	91.3	92.8	93.7
All classes	72.2	73.8	74.6	74.8	75.9

From the comparison between the baseline network, we can see that after using the light Swin Transformer backbone network to replace the original backbone layer, the ability to reshape features and extract local features can be strengthened, and the detection performance of small fabric defects can be improved. Compared with FPN, the weighted bidirectional feature network can enhance the feature fusion between different scales, improve the reuse of features, and facilitate the detection of small fabric defects. On the second basis, by adding a detection layer for comparison, both the map and recall rates are increased, indicating that the new detection layer improves the detection effect of small pixel defects. Comparing the results by YOLOV5 and YOLO-TBL, the recall rate is greatly improved after adding the generalized focal loss function, and the map also increases. The network is fully trained on the positive samples, which alleviates the extremely imbalanced number of background samples and foreground samples. Finally, the improved method proposed has an mAP of 75.9% and a recall rate of 73.1% in fabric defect detection. The experiments show that the improved algorithm can achieve high-precision and high-efficiency fabric defect detection.

In addition, we use a gradient-free algorithm Eigen CAM [31] to generate a network activation heat map to visually show the effect of improving the ablation. The comparison of the CAM heat map of the ablation experiment results in this section is shown in Figure 12.

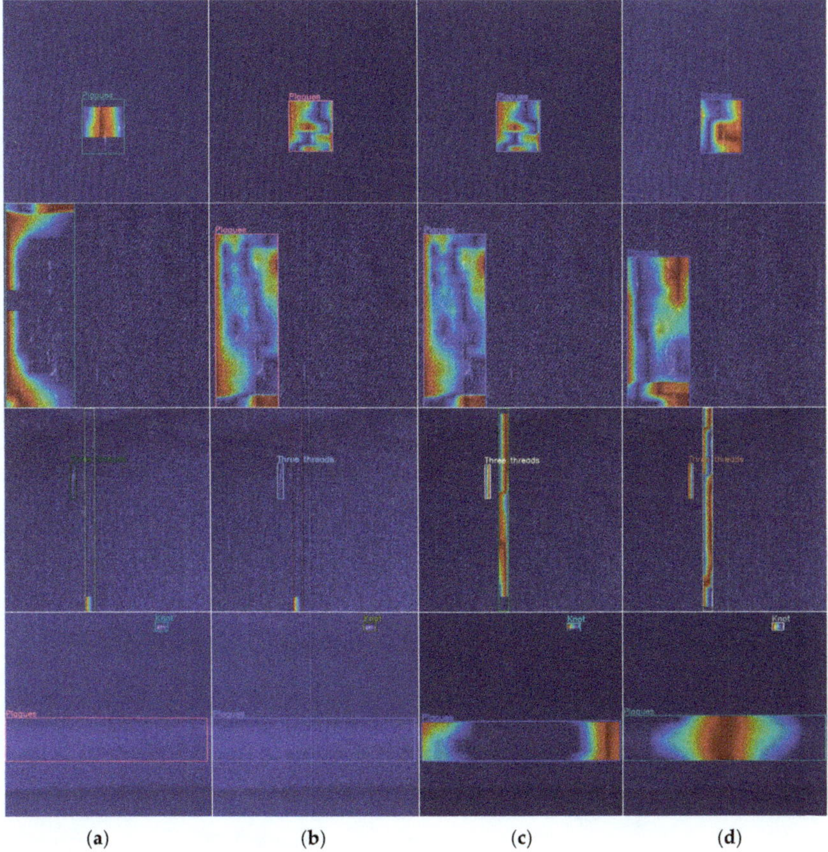

(a)   (b)   (c)   (d)

Figure 12. Heatmaps of different networks in ablation experiments: (**a**) CAM heat map of YOLO-LB; (**b**) CAM heat map of YOLO-SL; (**c**) CAM heat map of YOLO-SB; (**d**) CAM heat map of YOLO-TLB.

3.6. Ablation Experiments with Different Attention Mechanisms

To verify the improvement effect of the moving window-based mechanism proposed in this paper on the fusion of the YOLOv5 network model and other attention mechanisms, this paper adds several classic attention mechanisms to the neck layer for comparative experiments. Table 5 shows the experimental results of different attention. Figure 13 shows the effect of the CAM heatmap after adding the attention mechanism.

Table 5. Results of different attention mechanisms.

Attention Mechanism Model	mAP (%)
YOLO-TLB	75.9
+CBAM [32]	76.1 (+0.2)
+SE [33]	75.1 (−0.8)
+GAM [34]	76.5 (+0.6)

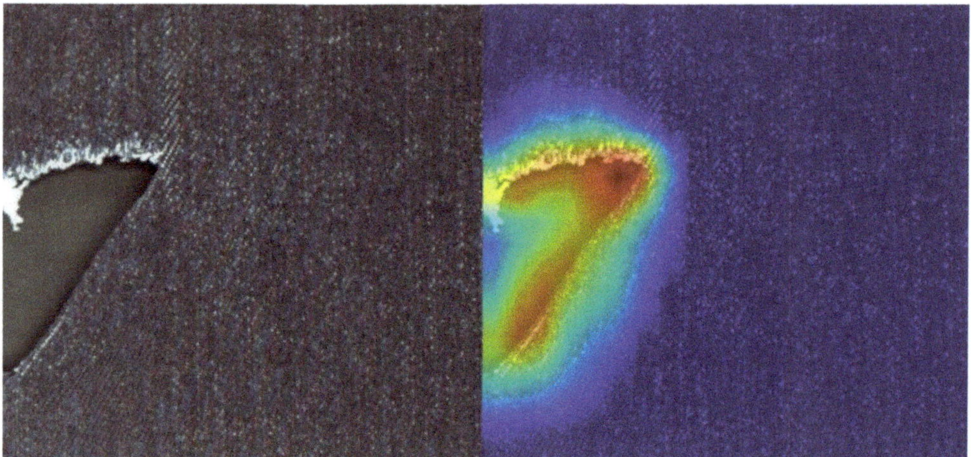

Figure 13. Attention of Eigen CAM heatmap.

This attention mechanism can explicitly show the area that the model pays attention to focus on. From the experimental results, it can be seen that after adding the Transformer's self-focus mechanism, mAP has increased, which can effectively strengthen the feature extraction capability of the model for low-resolution images, help to retain the feature information of fabric defects, and strengthen the network's ability to detect features around small defects. After introducing the GAM attention to the neck, the extraction ability of small objects is further improved.

3.7. Results Visualization

To more intuitively feel the detection effect of the modified method on defects, we choose several fabric images to compare the detection results. The left picture is the marked image, the middle picture is the unimproved YOLO algorithm detection picture, and the right picture is the modified algorithm detection picture. From Figure 14, the original YOLO algorithm to detect fabric defects has problems such as false detection of similar objects, missed detection of small defect objects, or poor detection effect of overlapping defects. The improved algorithm has a significantly improved detection effect.

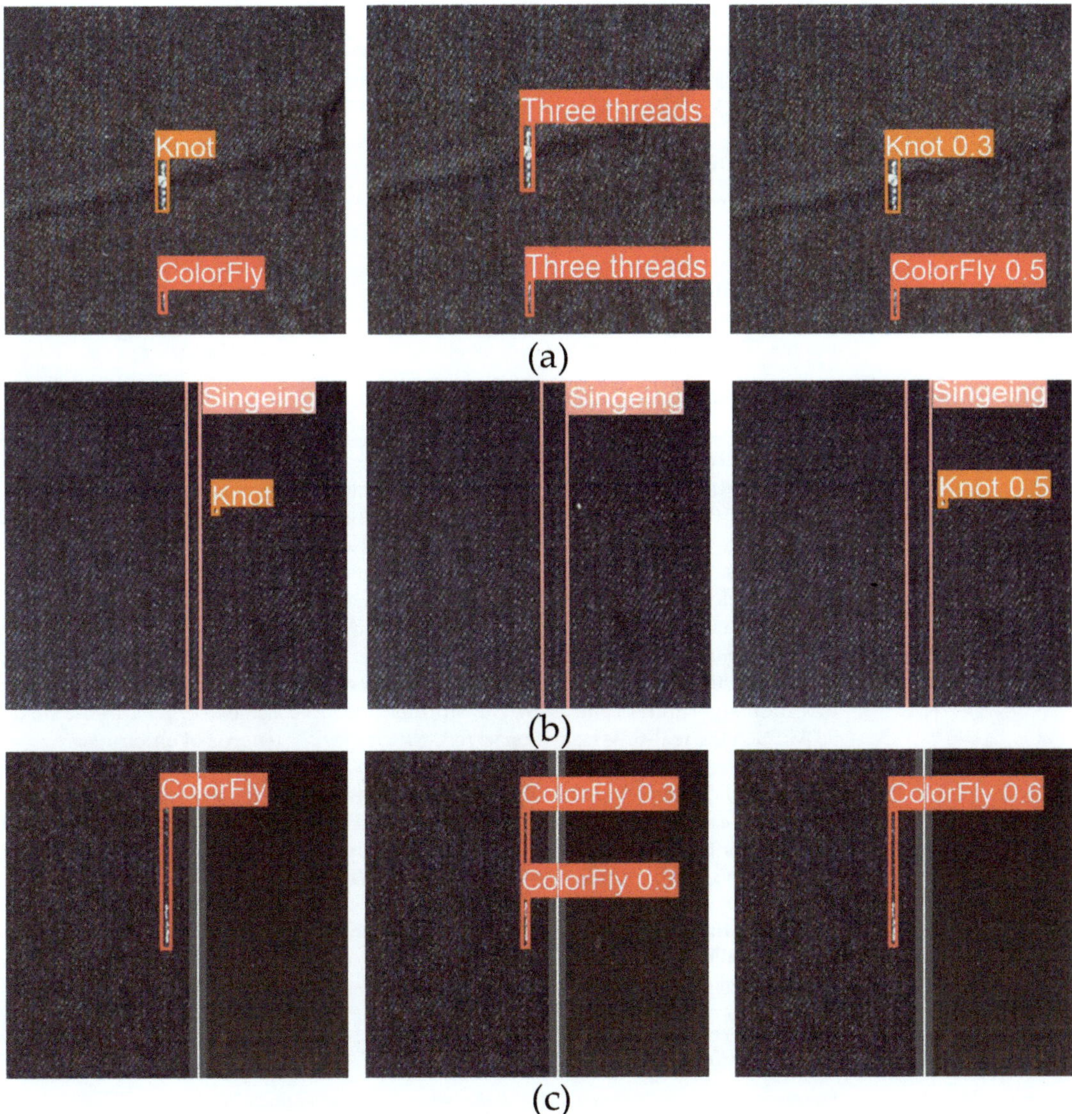

Figure 14. Some visual detection results. The first picture on the left is the labeled picture, the middle picture is the detection result of the original YOLOv5 algorithm, and the picture on the right is the detection result of the improved algorithm: (**a**) Detection result of the similar appearance of two kinds of fabric defects; (**b**) detection result of small target size defects; (**c**) detection result of overlapping fabric defects.

3.8. Comparison with State-of-the-Art Methods

To further demonstrate the advantages of the modified model in this paper, different mainstream target detection networks are trained under the same dataset, and the method proposed in our paper is compared to five classical target detection algorithms under the same experiment platform and dataset. The experimental results are shown in Table 6.

Table 6. Comparison with state-of-the-art methods.

Algorithm	Backbone Network	Precision (%)	mAP (%)	FPS
SDD [35]	VGG16	50.6	40.2	83.3
Faster R-CNN [36]	ResNet50	76.2	65.9	12.5
YOLOv3-Tiny [37]	Darknet53	46.1	46.7	113.6
YOLOv4-Mish [38]	CSPDarknet	67.8	58.8	111.1
YOLOv5 [39]	CSPDarknet	77.0	72.1	90.9
OUR	Swin Transformer	85.6	76.5	58.8

Through the comparison of the test results of various classic target networks on the fabric defect data set, the improved Swin Transformer algorithm proposed in this paper achieves an average accuracy of mAP of 76.5% and an accuracy of 85.6%. Compared with other detection methods, the algorithm has certain advantages, and the number of detections per second reaches 58.8, which meets the needs of enterprises and the actual situation.

4. Conclusions

This paper takes fabric defects as the research object, and realizes the precise location and classification of fabric defects by improving the algorithm. The YOLOv5 model is used as the baseline network, and the Swin Transformer encoder is added to the backbone network of the fabric defect detection model. The multiwindow sliding self-attention is added to strengthen the weighted bidirectional feature network, and a detection layer that can detect 4×4 small targets is added to the four-scale detection to enhance the detection of small defects around the extraction of feature information. The generalized focal loss function is introduced to strengthen the algorithmic learning of positive sample instances and reduce the missed detection rate. After experimental comparison with the above five target detection algorithms, compared with the original frame, the modified fabric defect detection algorithm of the Swin Transformer proposed achieves 76.5% in the mAP and 58.8 FPS in the real-time detection speed, which meets the needs of enterprises.

In subsequent research, we will continue to optimize and improve the fabric defect method. Under the condition of keeping the detection accuracy basically unchanged, the number of model calculation parameters and time consumption or the fabric samples required for training should be reduced so that the model can achieve the same detection effect in embedded or mobile terminals.

Author Contributions: Conceptualization, G.L.; methodology, G.L. and K.L.; software, G.L. and K.L.; validation, G.L., X.X. and K.L.; formal analysis, G.L. and K.L.; investigation, G.L. and K.L.; resources, G.L. and X.X.; data curation, K.L. and R.Y.; writing—original draft preparation, G.L. and K.L.; writing—review and editing, G.L. and K.L.; visualization, G.L. and K.L.; supervision, G.L.; project administration, G.L.; funding acquisition, G.L. All authors have read and agreed to the published version of the manuscript.

Funding: This research was funded by the Natural Science Foundation of China (No.51207133) and the Natural Science Foundation of Fujian Province, China (No.2020J01272).

Institutional Review Board Statement: Not applicable.

Informed Consent Statement: Not applicable.

Data Availability Statement: Not applicable.

Conflicts of Interest: The authors declare no conflict of interest.

References

1. Lu, W.; Lin, Q.; Zhong, J.; Wang, C.; Xu, W. Research progress of image processing technology for fabric defect detection. *Fangzhi Xuebao/J. Text. Res.* **2021**, *42*, 197–206. [CrossRef]
2. Gustian, D.A.; Rohmah, N.L.; Shidik, G.F.; Fanani, A.Z.; Pramunendar, R.A.; Pujiono. Classification of Troso Fabric Using SVM-RBF Multi-class Method with GLCM and PCA Feature Extraction. In Proceedings of the 2019 International Seminar on Application for Technology of Information and Communication, iSemantic 2019, Semarang, Indonesia, 21–22 September 2019; pp. 7–11.

3. Li, C.; Gao, G.; Liu, Z.; Huang, D.; Xi, J. Defect Detection for Patterned Fabric Images Based on GHOG and Low-Rank Decomposition. *IEEE Access* **2019**, *7*, 83962–83973. [CrossRef]

4. Pan, Z.; He, N.; Jiao, Z. FFT used for fabric defect detection based on CUDA. In Proceedings of the 2nd IEEE Advanced Information Technology, Electronic and Automation Control Conference, IAEAC 2017, Chongqing, China, 25–26 March 2017; pp. 2104–2107.

5. Wen, Z.; Cao, J.; Liu, X.; Ying, S. Fabric Defects Detection using Adaptive Wavelets. *Int. J. Cloth. Sci. Technol.* **2014**, *26*, 202–211. [CrossRef]

6. Chen, M.; Yu, L.; Zhi, C.; Sun, R.; Zhu, S.; Gao, Z.; Ke, Z.; Zhu, M.; Zhang, Y. Improved faster R-CNN for fabric defect detection based on Gabor filter with Genetic Algorithm optimization. *Comput. Ind.* **2022**, *134*, 103551. [CrossRef]

7. Yapi, D.; Mejri, M.; Allili, M.S.; Baaziz, N. A learning-based approach for automatic defect detection in textile images. *IFAC-PapersOnLine* **2015**, *48*, 2423–2428. [CrossRef]

8. Li, Y.; Zhao, W.; Pan, J. Deformable patterned fabric defect detection with fisher criterion-based deep learning. *IEEE Trans. Autom. Sci. Eng.* **2016**, *14*, 1256–1264. [CrossRef]

9. Zhang, H.-W.; Zhang, L.-J.; Li, P.-F.; Gu, D. Yarn-dyed fabric defect detection with YOLOV2 based on deep convolution neural networks. In Proceedings of the 2018 IEEE 7th Data Driven Control and Learning Systems Conference (DDCLS), Enshi, China, 25–27 May 2018; pp. 170–174.

10. Wang, Z.; Jing, J. Pixel-wise fabric defect detection by CNNs without labeled training data. *IEEE Access* **2020**, *8*, 161317–161325. [CrossRef]

11. Mei, S.; Wang, Y.; Wen, G. Automatic fabric defect detection with a multi-scale convolutional denoising autoencoder network model. *Sensors* **2018**, *18*, 1064. [CrossRef]

12. Huang, Y.; Jing, J.; Wang, Z. Fabric defect segmentation method based on deep learning. *IEEE Trans. Instrum. Meas.* **2021**, *70*, 1–15. [CrossRef]

13. Han, K.; Wang, Y.; Tian, Q.; Guo, J.; Xu, C.; Xu, C. Ghostnet: More features from cheap operations. In Proceedings of the IEEE/CVF Conference on Computer Vision and Pattern Recognition, Seattle, WA, USA, 13–19 June 2020; pp. 1580–1589.

14. Girshick, R.; Donahue, J.; Darrell, T.; Malik, J. Rich feature hierarchies for accurate object detection and semantic segmentation. In Proceedings of the IEEE Conference on Computer Vision and Pattern Recognition, Columbus, OH, USA, 23–28 June 2014; pp. 580–587.

15. Tan, M.; Pang, R.; Le, Q.V. Efficientdet: Scalable and efficient object detection. In Proceedings of the IEEE/CVF Conference on Computer Vision and Pattern Recognition, Seattle, WA, USA, 13–19 June 2020; pp. 10781–10790.

16. Santos, C.; Aguiar, M.; Welfer, D.; Belloni, B. A New Approach for Detecting Fundus Lesions Using Image Processing and Deep Neural Network Architecture Based on YOLO Model. *Sensors* **2022**, *22*, 6441. [CrossRef]

17. Chollet, F. Xception: Deep learning with depthwise separable convolutions. In Proceedings of the IEEE Conference on Computer Vision and Pattern Recognition, Honolulu, HI, USA, 21–26 July 2017; pp. 1251–1258.

18. Liu, H.; Sun, F.; Gu, J.; Deng, L. SF-YOLOv5: A Lightweight Small Object Detection Algorithm Based on Improved Feature Fusion Mode. *Sensors* **2022**, *22*, 5817. [CrossRef] [PubMed]

19. Redmon, J.; Divvala, S.; Girshick, R.; Farhadi, A. You only look once: Unified, real-time object detection. In Proceedings of the IEEE Conference on Computer Vision and Pattern Recognition, Las Vegas, NV, USA, 27–30 June 2016; pp. 779–788.

20. Redmon, J.; Farhadi, A. Yolov3: An incremental improvement. *arXiv* **2018**, arXiv:1804.02767.

21. Bochkovskiy, A.; Wang, C.-Y.; Liao, H.-Y.M. Yolov4: Optimal speed and accuracy of object detection. *arXiv* **2020**, arXiv:2004.10934.

22. Lin, T.-Y.; Dollár, P.; Girshick, R.; He, K.; Hariharan, B.; Belongie, S. Feature pyramid networks for object detection. In Proceedings of the IEEE Conference on Computer Vision and Pattern Recognition, Honolulu, HI, USA, 21–26 July 2017; pp. 2117–2125.

23. Rezatofighi, H.; Tsoi, N.; Gwak, J.; Sadeghian, A.; Reid, I.; Savarese, S. Generalized intersection over union: A metric and a loss for bounding box regression. In Proceedings of the IEEE/CVF Conference on Computer Vision and Pattern Recognition, Long Beach, CA, USA, 16–17 June 2019; pp. 658–666.

24. He, K.; Zhang, X.; Ren, S.; Sun, J. Spatial pyramid pooling in deep convolutional networks for visual recognition. *IEEE Trans. Pattern Anal. Mach. Intell.* **2015**, *37*, 1904–1916. [CrossRef] [PubMed]

25. Liu, Z.; Lin, Y.; Cao, Y.; Hu, H.; Wei, Y.; Zhang, Z.; Lin, S.; Guo, B. Swin transformer: Hierarchical vision transformer using shifted windows. In Proceedings of the IEEE/CVF International Conference on Computer Vision, Nashville, TN, USA, 20–25 June 2021; pp. 10012–10022.

26. Shaw, P.; Uszkoreit, J.; Vaswani, A. Self-attention with relative position representations. *arXiv* **2018**, arXiv:1803.02155.

27. Chen, J.; Mai, H.; Luo, L.; Chen, X.; Wu, K. Effective feature fusion network in BIFPN for small object detection. In Proceedings of the 2021 IEEE International Conference on Image Processing (ICIP), Anchorage, AK, USA, 19–22 September 2021; pp. 699–703.

28. Lin, T.-Y.; Goyal, P.; Girshick, R.; He, K.; Dollár, P. Focal loss for dense object detection. In Proceedings of the IEEE International Conference on Computer Vision, Venice, Italy, 22–29 October 2017; pp. 2980–2988.

29. Li, X.; Wang, W.; Wu, L.; Chen, S.; Hu, X.; Li, J.; Tang, J.; Yang, J. Generalized focal loss: Learning qualified and distributed bounding boxes for dense object detection. *Adv. Neural Inf. Process. Syst.* **2020**, *33*, 21002–21012.

30. Yu, X.; Lyu, W.; Zhou, D.; Wang, C.; Xu, W. ES-Net: Efficient Scale-Aware Network for Tiny Defect Detection. *IEEE Trans. Instrum. Meas.* **2022**, *71*, 1–14. [CrossRef]

31. Selvaraju, R.R.; Cogswell, M.; Das, A.; Vedantam, R.; Parikh, D.; Batra, D. Grad-cam: Visual explanations from deep networks via gradient-based localization. In Proceedings of the IEEE International Conference on Computer Vision, Venice, Italy, 22–29 October 2017; pp. 618–626.

32. Woo, S.; Park, J.; Lee, J.-Y.; Kweon, I.S. Cbam: Convolutional block attention module. In Proceedings of the European Conference on Computer Vision (ECCV), Munich, Germany, 8–14 September 2018; pp. 3–19.
33. Hu, J.; Shen, L.; Sun, G. Squeeze-and-excitation networks. In Proceedings of the IEEE Conference on Computer Vision and Pattern Recognition, Salt Lake City, UT, USA, 8–12 June 2018; pp. 7132–7141.
34. Liu, Y.; Shao, Z.; Hoffmann, N. Global Attention Mechanism: Retain Information to Enhance Channel-Spatial Interactions. *arXiv* **2021**, arXiv:2112.05561.
35. Liu, X.; Gao, J. Surface Defect Detection Method of Hot Rolling Strip Based on Improved SSD Model. In Proceedings of the International Conference on Database Systems for Advanced Applications, Taipei, Taiwan, 11–14 April 2021; pp. 209–222.
36. Zhao, W.; Huang, H.; Li, D.; Chen, F.; Cheng, W. Pointer defect detection based on transfer learning and improved cascade-RCNN. *Sensors* **2020**, *20*, 4939. [CrossRef]
37. Sujee, R.; Shanthosh, D.; Sudharsun, L. Fabric Defect Detection Using YOLOv2 and YOLO v3 Tiny. In Proceedings of the International Conference on Computational Intelligence in Data Science, Chennai, India, 20–22 February 2020; pp. 196–204.
38. Dlamini, S.; Kao, C.-Y.; Su, S.-L.; Jeffrey Kuo, C.-F. Development of a real-time machine vision system for functional textile fabric defect detection using a deep YOLOv4 model. *Text. Res. J.* **2022**, *92*, 675–690. [CrossRef]
39. Jin, R.; Niu, Q. Automatic Fabric Defect Detection Based on an Improved YOLOv5. *Math. Probl. Eng.* **2021**, *2021*, 7321394. [CrossRef]

Article

Research on State Recognition Technology of Elevator Traction Machine Based on Modulation Feature Extraction

Dongyang Li [1,2], Jianyi Yang [1,*] and Yong Liu [3]

1   College of Information Science and Electronic Engineering, Zhejiang University, Hangzhou 310013, China
2   Hangzhou Special Equipment Inspection and Research Institute, Hangzhou 310051, China
3   College of Control Science and Engineering, Zhejiang University, Hangzhou 310013, China
*   Correspondence: yangjy@zju.edu.cn

Abstract: AbstractVibration signal analysis of the traction machine is an important part of the current rotating machinery state recognition technology, and its feature extraction is the most critical step. In this study, the time-frequency characteristics of the vibration of the traction machine under different elevator running directions, running speeds and load weights are analyzed. The novel demodulation method based on time-frequency analysis and principal component analysis (DPCA) is used to extract the periodic modulated wave signal. In order to compare different influence of background noise and unknown frequency influence, the Fast Fourier Transform (FFT) and Short Time Fourier Transform (STFT) methods are used to extract the characteristics of the traction machine vibration signal, respectively. Under different load conditions, it is difficult to observe the obvious differences and similarities of the vibration signals of the traction machine by time-frequency method. However, the DPCA demodulation method provides a guarantee for the reliability and accuracy of the state identification of the traction machine.

Keywords: vibration signal; traction machine; feature extraction; state identification

Citation: Li, D.; Yang, J.; Liu, Y. Research on State Recognition Technology of Elevator Traction Machine Based on Modulation Feature Extraction. *Sensors* **2022**, *22*, 9247. https://doi.org/10.3390/s22239247

Academic Editors: Yongbo Li and Ruqiang Yan

Received: 6 October 2022
Accepted: 24 November 2022
Published: 28 November 2022

Publisher's Note: MDPI stays neutral with regard to jurisdictional claims in published maps and institutional affiliations.

1. Introduction

Elevator is a large-scale complex equipment integrating mechanical, electrical and control. If it breaks down, it will directly affect the safe and efficient operation of the elevator. Nowadays, elevators are used more and more frequently in daily life and production. Therefore, the number of elevators shows a trend of continuous growth [1], and the accompanying elevator failure and maintenance problems are becoming more and more prominent [2,3]. As the power device of the elevator, the traction machine determines whether the elevator can operate normally. With the vigorous development of science and technology, the state recognition technology of elevator traction machine is also constantly improving [4,5].

In recent years, the fault diagnosis technology of elevator traction machine with the artificial intelligence [6] or image processing [7] has been widely applied and developed. However, these methods face problems, such as non-universality of diagnostic model, high cost of model training, and requirement for massive fault samples. In addition, the selection of fault features is also of great significance to the optimization of diagnosis model.

Traction machine is a complex mechanical structure, which is closely connected by various parts. Therefore, the state identification of the traction machine can be diagnosed by various signals, such as vibration, noise, current, temperature, braking torque, speed, and power. Many useful information is hidden in the vibration signal of the traction machine [8]. These signal characteristics can reflect the working condition of the equipment. By analyzing the vibration characteristics of the equipment, the safety operation, accident prevention and maintenance cost reduction can all be accomplished.

Based on the vibration signals, a lot of research have been done in which signal feature extraction methods are the most important section of fault diagnosis [9,10]. The time domain method and frequency domain method have been commonly used in early fault diagnosis engineering [11]. Filtering, amplification, statistical feature calculation, correlation analysis, and other time-domain signal processing are referred to as time-domain signal analysis. However, it only reflects the change of amplitude with time and lack frequency bands information. The frequency domain analysis method is to describe the raw signal in the frequency domain, which is more intuitive than the time domain analysis method. However, traditional frequency domain analysis might fail to extract the characteristics information of traction machines due to the heavy background noise and complicated excitation sources [12]. Therefore, only relying on the frequency domain analysis method is far from meeting the current requirements of traction machine fault diagnosis. This brings huge challenges to the status identification and fault diagnosis of traction machines. Therefore, a time-frequency combination processing method has been proposed. Short-time Fourier transform (STFT) [13,14] and wavelet transform (WT) [15] are the common used processing tools with fine time localization and frequency resolution. These methods are realized by superposition of Fourier transform in different fixed window length. However, due to lacking self-adaptability, the quality of feature extraction might be affected by the selection of window function or wavelet basis function.

In addition, the above methods can extract the characteristics of vibration signals, but the collected vibration signals usually contain background noise and unknown frequency interference. To eliminate noise component and extract the fault feature information of raw vibration signals, several demodulation techniques have been applied to past research, such as Hilbert transform (HT) [16], empirical mode composition (EMD) [17], spectral kurtosis (SK) [18], nonstationary analysis [19–21], and cyclostationary analysis [22–24]. These methods have been applied to modulation frequency extraction already, which noted the modulation mechanism in a rotating machine.

Feng et al. [25,26] proposed an adaptive iterative generalized demodulation method to extract the modulation features in nonstationary analysis. The vibration characteristics of hydraulic turbine and planetary gearbox have been successfully found in the joint time-frequency domain. Most vibration signals of traction machine are non-stationary signals, but they are cyclostationary signals, namely, the correlation function of traction machine signals is periodic function of time. In view of the cyclostationary analysis theory, a variety of methodologies have been proposed, in which cyclic modulation spectrum (CMS) and fast spectral correlation (Fast-SC) are two typical cyclostationary tools [22]. However, they did not gain its deserved attention because of high computational cost.

Wang et al. [27] improved the cyclostationary methods with an application of Teager Kaiser energy operator (TKEO), which can enhance fault feature recognition with low computational burden. Song et al. [28,29] proposed a demodulation method based on time-frequency analysis (TSA) and principal component analysis (PCA) and applied it to the modulation frequency extraction of pump and permanent magnet synchronous motor (PMSM). Moreover, due to dimensionality reduction of time-frequency distribution matrix, the burden of high computational cost was greatly relieved. The main process of the algorithm is as follows: Firstly, the raw vibration signal is transformed into time-frequency domain by STFT. Then, the PCA method is used to reduce the dimensionality of the time-frequency spectrum in order to extract the eigenvalues of the principal components. Finally, the principal components are reconstructed to obtain the modulation signals.

Among the above demodulation methods, it could be found that the demodulation method base on PCA (DPCA) has great potential for applications in traction machine. In addition, although the fault diagnosis technology of elevator traction machine based on artificial intelligence or image processing has been widely applied and developed. However, few investigations have been done to extract and analyze the modulation features of traction. The modulation mechanism of traction machine has also rarely been involved. These above issues have greatly hindered the development of elevator fault diagnosis technology.

In this paper, the modulation characteristics of the traction machine vibration signal were extracted through a demodulation method based on time-frequency analysis and principal component analysis (DPCA). The characteristics extracted by DPCA is more prominent under the interference of background noise and unknown frequency, which is helpful to the state identification of the traction machine. The principle of signal demodulation method and experiential setting are introduced respectively in Section 2. In Section 3, the vibration signal of the traction machine is processed by FFT, STFT and DPCA methods. The influence of different working conditions on the vibration of traction machine is discussed, which shows the superiority of the demodulation technology. Finally, the conclusions are drawn in Section 4.

2. Methodology

2.1. DPCA Method

To identify the state of machinery, the mainly three steps are as follows: acquisition of monitoring signals, feature extraction of monitoring signals, and pattern recognition and diagnosis of the state are carried out. For the traction machine state recognition technology, the extraction of state features is hard work, which directly affects the accuracy of state diagnosis and the reliability of early prediction.

The state parameters of the tractor during operation are hidden in the raw signals. Therefore, the extraction of the state parameters has become an important factor affecting the accuracy of the state identification. Based on the feature extraction, various signal processing techniques have been developed, which mainly involved time domain analysis, frequency domain analysis, time-frequency analysis, etc. [30]. Although the above methods can extract the features of vibration signals, the collected vibration signals usually contain background noise and unknown frequency interference. The amplitude demodulation process (also known as high frequency resonance, resonance demodulation or envelope analysis) separates low frequency from high frequency background noise [31]. In this paper, the DPCA method was adopted for feature extraction [28]. DPCA algorithm mainly includes: time-frequency analysis, principal component analysis and feature extraction.

(1) Time frequency analysis.

When the traction machine operates stably, its key modulation component is modulation signal. The single component modulation signal of the traction machine can be expressed in Equation (1), which is mainly composed of modulation signal and carrier signal.

$$x(t) = x_m(t)x_c(t) \tag{1}$$

where $x(t)$ is the amplitude modulation signal of the traction machine, $x_m(t)$ is the modulation signal, and $x_c(t)$ is the carrier signal.

The time-frequency distribution of the monitoring signal can be expressed in Equation (2).

$$P_X(f,t) = \int_{-\infty}^{\infty} x_m(\tau)x_c(\tau)w(t-\tau)e^{-j2\pi f\tau}d\tau \tag{2}$$

where $P_X(f, t)$ is the time-frequency distribution function of the monitoring signal, and $w(t)$ is the window function of the STFT.

The STFT of the modulation signal model of the traction machine can be approximated as follows, as shown in Equation (3).

$$\int_{-\infty}^{\infty} x_m(\tau)x_c w(t-\tau)e^{-j2\pi f\tau}d\tau \approx x_m(\tau)\int_{-\infty}^{\infty} x_c(\tau)w(t-\tau)e^{-j2\pi f\tau}d\tau \tag{3}$$

The time spectrum of the modulated signal is further simplified to obtain Equation (4).

$$P(f,t) \approx x_m(t)\int_{-\infty}^{\infty} x_c(\tau)w(t-\tau)e^{-j2\pi f\tau}d\tau = x_m(t)P_C(f,t) \tag{4}$$

where $P(f, t)$ represents the time-frequency distribution function of the detection signal, $P_C(f, t)$ represents the time-frequency distribution function of the carrier signal.

(2) Principal component analysis.

Principal component analysis is a classical data dimensionality reduction method, which is mainly realized by the following algorithms.

Firstly, the covariance matrix is solved. The matrix formula is shown in Equation (5).

$$P_{\text{cov}} = \text{cov}(P(t, f)) \tag{5}$$

where P_{cov} represents the covariance matrix of the time-frequency distribution matrix, cov() represents the covariance operator.

Secondly, there is eigenvalue decomposition. As shown in Equation (6).

$$[\mathbf{V}, \mathbf{U}] = \text{eig}(P_{\text{cov}}) \tag{6}$$

where eig() represents the eigenvalue decomposition operator. $\mathbf{V}$, $\mathbf{U}$ represent eigenvalue matrix and eigenvector matrix respectively.

Thirdly, eigenvalue selection. The order of the selected eigenvalue is determined by the maximum value of the difference spectrum, as shown in Equation (7).

$$k \geq i\big|_{\max(\delta_i = (\lambda_i - \lambda_{i+1}))} \tag{7}$$

where k represents the order of the selected eigenvalue, δ_i represents the difference spectrum value.

Finally, principal component reconstruction. The corresponding principal component modulation signal $\text{PPC}_i(t)$ can be obtained, as shown in Equation (8).

$$\text{PPC}_i(t) = P(t, f)u_i \tag{8}$$

(3) Feature extraction.

The principal component analysis method can be used to obtain the principal component of the monitoring signal, which includes the low-frequency modulation component of the monitoring signal. The characteristic modulation frequency can be extracted by frequency analysis, as shown in Equation (9).

$$P_i(f) = \int_{-\infty}^{\infty} \text{PPC}_i(t)e^{-j2\pi ft}dt \tag{9}$$

2.2. Elevator Traction Machine Parameters

The model of tractor selected in the experiment is GETM3.DM. The detailed parameters are listed in Table 1.

Table 1. Parameters of elevator traction machine.

Parameter	Value
Model	GETM3.DM
Moment of inertia [kg·m^2]	4.4
Pulley diameter [mm]	400
Rated voltage [V]	513
Rated current [A]	12.6
Rated power [kW]	9.7
Rated speed [rpm]	168
Rated frequency [Hz]	28

2.3. Equipment Selection

The test system was used for the state identification research of elevator traction machine, as shown in Figure 1. The instruments included vibration acceleration sensor, data acquisition instrument, computer, and other auxiliary instruments. The acceleration sensor was fixed to the traction machine, and the vibration signals collected by the sensor were transmitted to the data acquisition instrument.

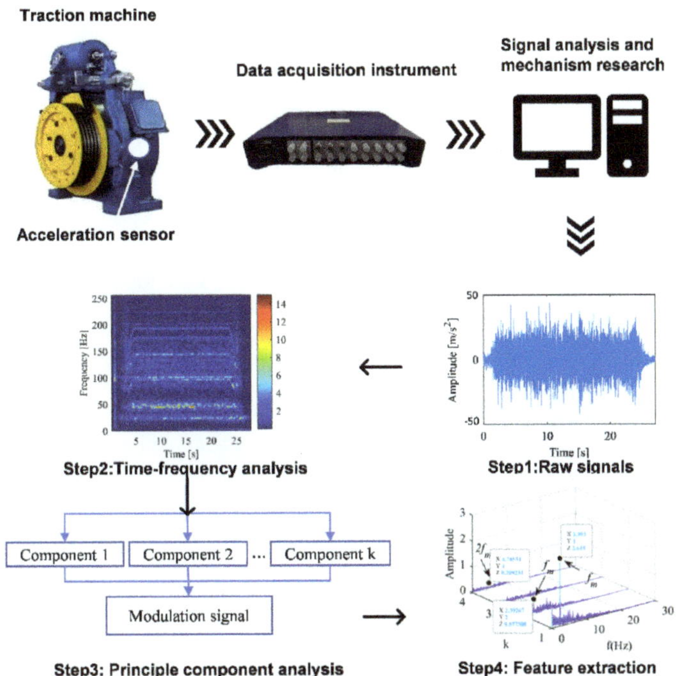

Figure 1. Flow chart of traction machine vibration signal acquisition and analysis.

2.4. Test Conditions

In order to realize the recognition of different states of the traction machine, the test conditions involved in this paper are listed in Table 2.

Table 2. Test conditions.

Working Condition	Classification		
Different running directions	(a) Elevator up	(b) Elevator down	
Different operating speeds	(a) 1 m/s	(b) 2.4 m/s	
Different loads	(a) no-load	(b) 140 kg	(c) 325 kg

3. Case Analysis

3.1. Analysis of Influence of Elevator Running Speed on Main Engine Vibration

To compare features extracted by the different signal analysis methods, FFT was used to transform the time domain signal into spectrum domain. Their peaks value of the spectrum under different conditions are recorded in Tables 3 and 4. Under the working condition of 1 m/s, the frequency spectrum, time-frequency spectrum, and DPCA result are shown in Figures 2–4, respectively. Under the working condition of 2.4 m/s, the frequency spectrum, time-frequency spectrum, and DPCA result are shown in Figures 5–7, respectively.

With the comparison of the FFT results, it can be found that operating speed has an impact on the amplitude of the vibration spectrums. The amplitude of each frequency under low-speed operation (1.0 m/s) was lower than the amplitude of each frequency under normal operation (2.4 m/s), as shown in Figures 2 and 5. In general, the peak frequency in vibration spectrum will increase with acceleration of the traction machine.

In the time-frequency spectrum shown in Figure 3, it can be found that the prominent frequency of the elevator machine at the operating speed of 1 m/s is approximately 25 Hz. While the prominent vibration frequency under normal speed (2.4 m/s) is approximately 50 Hz, as shown in Figure 6. This characteristic is positively related to the operating speed of the traction machine, which can be regard as the main feature of elevator vibration. The vibration level of the traction machine can be evaluated by the amplitude change of this frequency band in the time-frequency spectrum.

Through comparative analysis, the DCPA results shown in Figures 4 and 7a. The k refers to the serial number of principle frequency bands selected by Equation (7). The modulated frequency of elevator traction machine is indicated by f_m, which can be found in each modulation spectrum. It can be observed that when $k = 1$, the amplitude modulation difference of f_m under two working conditions is 2.58 times. This value is approximate to the speed ratio under two working conditions. Therefore, the mechanism of vibration level-up caused by the operating speed-up is the increase of modulation effect in principle frequency bands.

As a result, for the working conditions with obvious differences, such as the influence of different operating speeds of the elevator on the vibration signal of the traction machine, the difference between the two states could be obtained by analyzing the frequency-domain diagram through the FFT. The time-frequency diagram and the demodulation diagram can more clearly highlight the difference and complete the identification of the state of the traction machine.

Table 3. Frequency domain peak value of vibration response at running speed of 1 m/s.

Peak Sequence Number	1	2	3	4	5	6	7
Frequency (Hz)	24.1	64.2	96.3	100.8	117.2	130.0	201.6
Up-drive (m/s^2)	1.48	0.51	0.52	1.46	2.91	2.67	0.64
Down-drive (m/s^2)	1.46	0.49	0.36	0.79	2.51	3.06	0.82

Table 4. Frequency domain peak value of vibration response in different running directions of elevator.

Peak Sequence Number	1	2	3	4	5	6	7	8
Frequency (Hz)	24.1	48.1	96.3	100.8	144.5	175.8	192.6	223.1
Up-drive (m/s^2)	2.00	1.63	2.35	3.15	2.17	0.76	0.96	0.24
Down-drive (m/s^2)	1.85	3.60	2.17	3.33	0.66	0.53	0.80	0.25

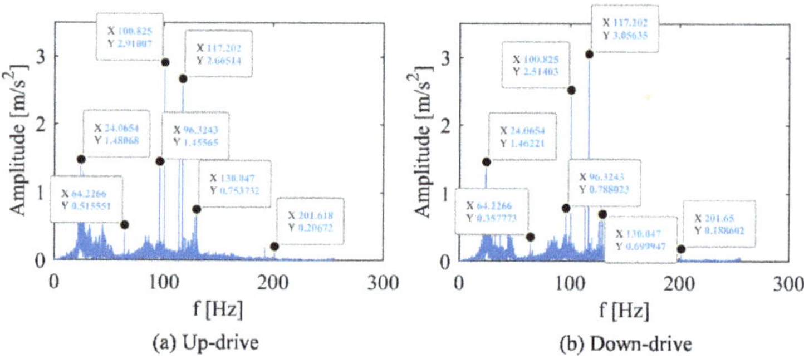

Figure 2. Vibration spectrum diagram of main engine at elevator running speed of 1 m/s. (**a**) Motor with up-drive condition. (**b**) Motor with down-drive condition.

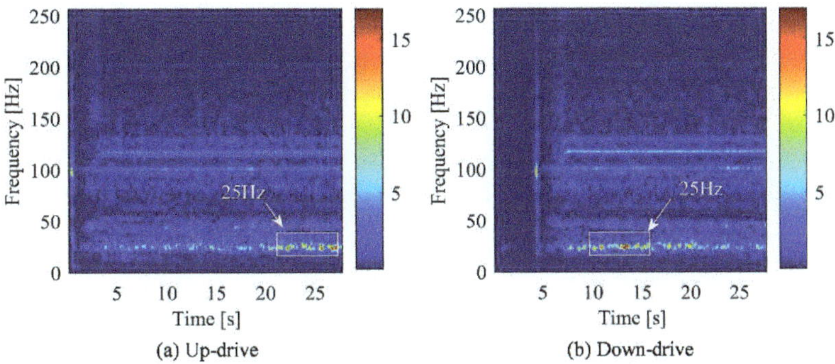

Figure 3. Time-frequency diagram of main engine vibration at elevator running speed of 1 m/s. (**a**) Motor with up-drive condition. (**b**) Motor with down-drive condition.

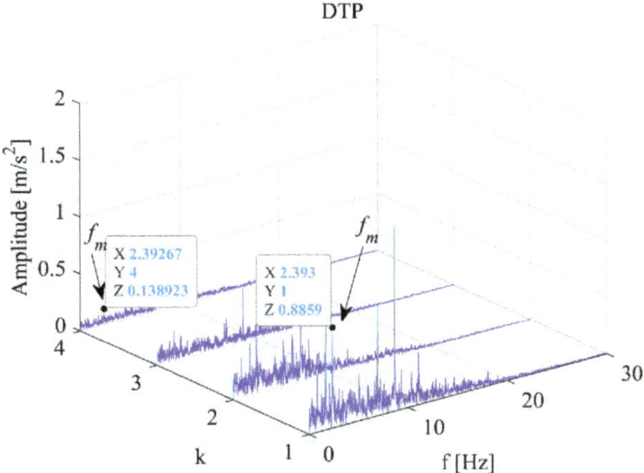

Figure 4. Vibration demodulation diagram of main engine when elevator speed is 1 m/s.

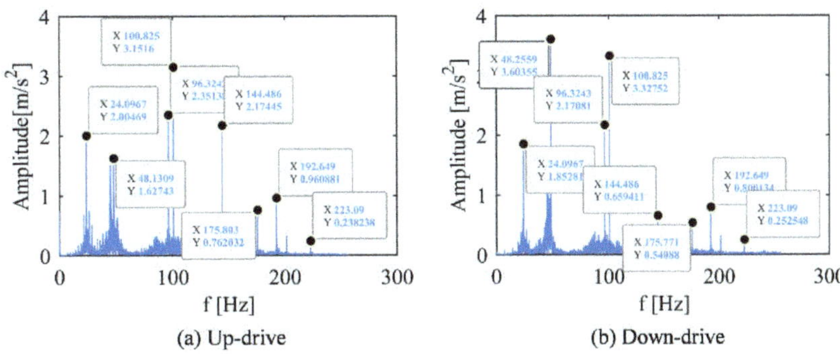

Figure 5. Vibration spectrum of main engine in different running directions of elevator (2.4 m/s). (**a**) Motor with up-drive condition. (**b**) Motor with down-drive condition.

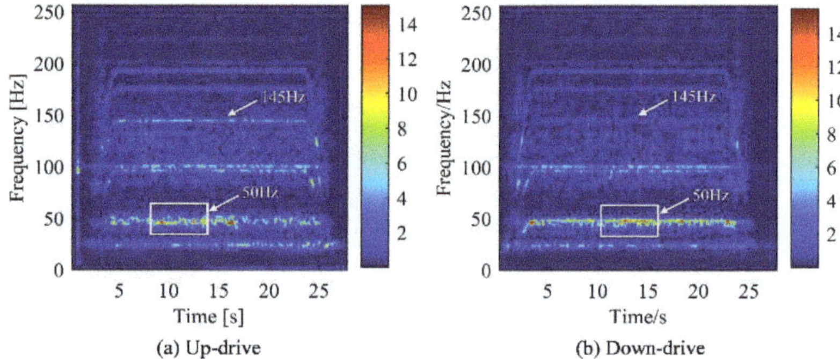

(a) Up-drive   (b) Down-drive

Figure 6. Time frequency diagram of main machine vibration in different operating directions of elevator (2.4 m/s). (**a**) Motor with up-drive condition. (**b**) Motor with down-drive condition.

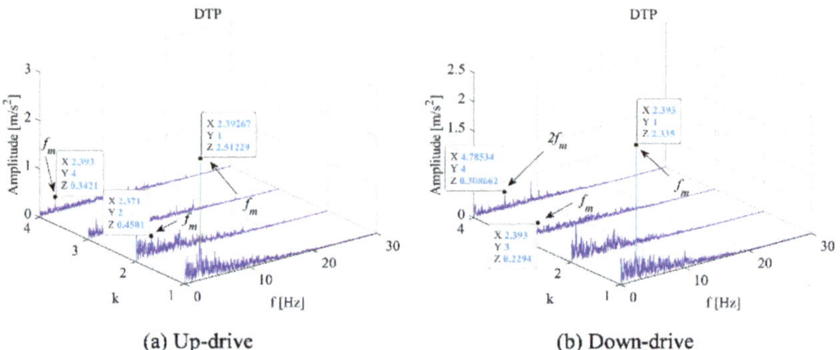

(a) Up-drive   (b) Down-drive

Figure 7. Vibration demodulation diagram under different operating conditions (2.4 m/s). (**a**) Motor with up-drive condition. (**b**) Motor with down-drive condition.

3.2. Analysis on the Influence of Elevator Running Direction on the Vibration of Main Engine

The spectrum analysis about different running direction is shown in Figure 5 and Table 4. It can be found that the vibration response was largest at 100.8 Hz under up-drive operation, while the largest vibration response under down-drive operation was at 48.1 Hz.

When the frequency was 48.1 Hz, the peak value of the downlink is much larger than that of the uplink, which was twice that of the downlink. As the frequency was 144.5 Hz, the peak value of the uplink was much larger than that of the downlink, which was three times that of the downlink. Except for 48.1 Hz and 144.5 Hz, the characteristic frequency of the most obvious peak under the two working conditions of the elevator was basically unchanged, and the height of the main peak slightly changed.

According to the analysis of Figure 6, the frequency (rotation speed) of the elevator gradually increased from the start to a certain state and then remains stable. After a cycle of operation, the frequency gradually decreased. Comparing (a) and (b) in Figure 6, it can be found that the peak value of the uplink was greater than that of the downlink at 145 Hz, while the difference in other frequency bands were not significant.

From the time-frequency spectrum shown in Figure 6, it can only be concluded that the difference between the two working conditions was the most significant at the frequency of 145 Hz. However, it was not enough to support the identification of the elevator's up and down conditions. Therefore, based on the spectrum analysis of STFT, the modulation signal in the vibration signal of the traction machine was extracted by the PCA technology.

The vibration demodulation diagram of the main engine under different operating conditions of the elevator in Figure 7 was analyzed. When $k = 1$, there was little difference between the up-working condition and the down-working condition; when $k = 2$, the frequency modulation of one f_m is generated more in the upstream working condition than in the downstream working condition; when $k = 3$, a f_m frequency modulation is generated in the downstream working condition more than in the upstream working condition; and when $k = 4$, the uplink has a frequency modulation of f_m and the downlink has a frequency modulation of $2f_m$.

For the up and down working conditions of the elevator, the influence on the traction machine was not obvious. Only the frequency domain diagram and the time-frequency diagram cannot accurately distinguish the two working conditions. Therefore, the demodulation method was used to separate the signal from the raw signal, highlight the weak state characteristic signal, and distinguish the states of different working conditions.

3.3. Analysis of Influence of Elevator Load on Main Engine Vibration

Taking the behavior under the elevator as an example, the measured time domain diagram was transformed into a spectrum diagram by FFT, as shown in Figure 8.

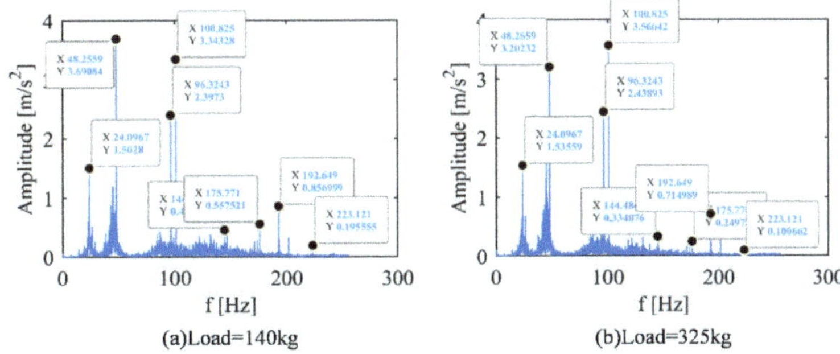

(a)Load=140kg   (b)Load=325kg

Figure 8. Vibration spectrum diagram of elevator main engine with different loads. (**a**) Motor running under 140 kg load condition. (**b**) Motor running under 325 kg load condition.

From the Figures 5b and 8 and Tables 4 and 5, it was hard to identify different working conditions only by using the frequency domain diagram. Then, the time-frequency spectrums obtained by STFT was analyzed, as shown in Figure 9.

Table 5. Frequency domain peak value of vibration response of elevator with different loads.

Peak Sequence Number	1	2	3	4	5	6	7	8
Frequency (Hz)	24.1	48.1	96.3	100.8	144.5	175.8	192.6	223.1
0 kg	1.85	3.60	2.17	3.33	0.66	0.53	0.80	0.25
140 kg	1.50	3.69	2.40	3.34	0.42	0.56	0.86	0.20
325 kg	1.54	3.20	2.44	3.57	0.33	0.25	0.71	0.28

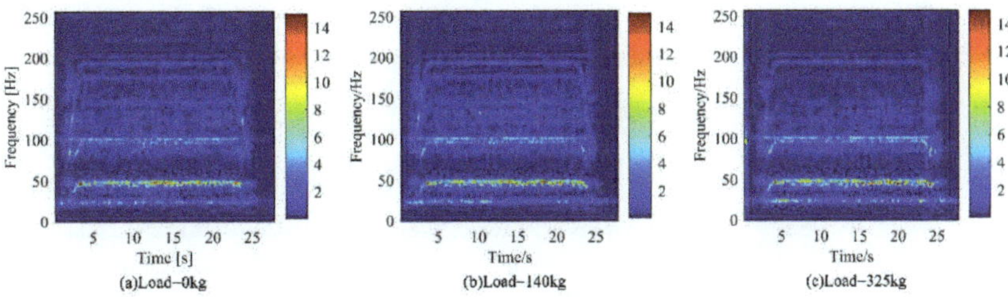

(a)Load−0kg   (b)Load−140kg   (c)Load−325kg

Figure 9. Vibration time-frequency diagram of down main machine under different loads of elevator. (**a**) Motor running under no-load condition. (**b**) Motor running under 140 kg load condition. (**c**) Motor running under 325 kg load condition.

However, it was also difficult to distinguish the working conditions of different loads by time-frequency spectrums. Therefore, based on the spectrum obtained by STFT, the modulation signal in the vibration signal of the traction machine was extracted by the DPCA method.

From the vibration demodulation diagram of the elevator main engine in Figure 10, it can be found that the frequency modulation of $2f_m$ was less than that of the other two working conditions when $k = 1$, and the frequency modulation of $2f_m$ was less than that of the other two directions when $k = 2$. When the load was 325 kg and $k = 3$, the frequency modulation of f_m was less than that of other working conditions. In addition, the amplitude corresponding to the common frequency modulation in the three working conditions increases with the increase of the load.

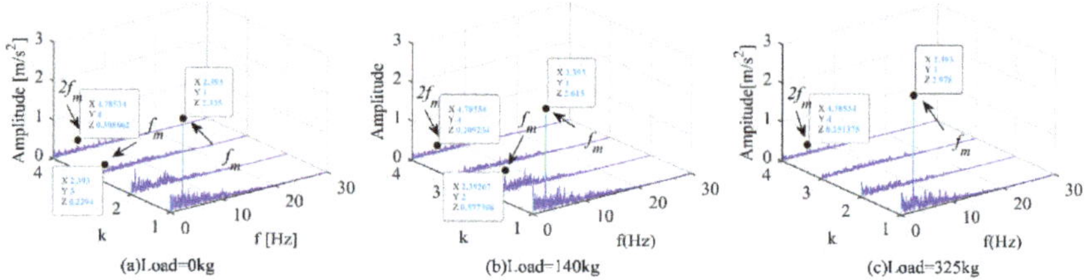

Figure 10. Vibration demodulation diagram of down main machine under different loads of elevator. (**a**) Motor running under no-load condition. (**b**) Motor running under 140 kg load condition. (**c**) Motor running under 325 kg load condition.

According to the working conditions of different loads of the elevator, the influence on the traction machine is not obvious. It was difficult to distinguish the two working conditions only through the frequency-domain diagram and the time-frequency diagram. Therefore, the demodulation method is used to separate the signal from the raw signal, highlight the weak state characteristic signal, and distinguish the states of different working conditions.

4. Conclusions

In this paper, the application of the DCPA method provides an alternative way to realize fast and effective condition monitoring of a traction machine, which could be extended to detect other background interference and typical faults. The conclusion is as follows:

(1) For the influence of different operating speeds of the elevator on the vibration signal, the difference can be obtained by analyzing the frequency-domain diagram through the FFT. The time-frequency diagram and the demodulation diagram can more clearly highlight the difference and complete the identification of the state of the traction machine. The amplitude modulation ratio of f_m is approximate to the speed ratio under working conditions for different speeds;

(2) For the up and down working conditions of the elevator, the frequency domain diagram and the time-frequency diagram cannot accurately distinguish the two working conditions. The DPCA demodulation method could highlight the weak state characteristic signal and distinguish the states of different working conditions;

(3) Under different load conditions, it is difficult to observe the obvious differences and similarities of the vibration signals of the traction machine by time-frequency method. However, the DPCA demodulation method can effectively solve the influence of background noise and unknown frequency interference of the traction machine vibration signal. With the increase of load, the amplitude modulation of shaft frequency (f_m) increases;

(4) The state identification technology discussed in this paper involved a healthy traction machine under various operation. The state identification of traction machines with different faults will be carried out in future work.

Author Contributions: Conceptualization, D.L. and J.Y.; methodology, D.L.; software, D.L.; validation, D.L., J.Y. and Y.L.; formal analysis, D.L.; investigation, D.L.; resources, D.L.; data curation, D.L.; writing original draft preparation, D.L.; writing—review and editing, D.L.; visualization, D.L.; supervision, D.L.; project administration, D.L.; funding acquisition, J.Y. All authors have read and agreed to the published version of the manuscript.

Funding: This research was funded by Open Research Project of the State Key Laboratory of Industrial Control Technology grant number [No. ICT2022B13]. The APC was funded by the Science and Technology Project of the State Administration for Market Regulation [No. 2021MK140].

Institutional Review Board Statement: Not applicable.

Informed Consent Statement: Not applicable.

Data Availability Statement: Data will be made available on request.

Acknowledgments: This work was supported by the Open Research Project of the State Key Laboratory of Industrial Control Technology, Zhejiang University, China (No. ICT2022B13) and the Science and Technology Project of the State Administration for Market Regulation (No. 2021MK140).

Conflicts of Interest: The authors declare no conflict of interest.

References

1. Cai, N.; Chow, W.K. Numerical Studies on Fire Hazards of Elevator Evacuation in Supertall Buildings. *Indoor Built Environ.* **2019**, *28*, 247–263. [CrossRef]
2. Wen, P.; Zhi, M.; Zhang, G.; Li, S. Fault Prediction of Elevator Door System Based on PSO-BP Neural Network. *Engineering* **2016**, *8*, 761–766. [CrossRef]
3. Liu, J.; Gong, Z.; Bai, Z.; Gu, M.; Liu, M.; Chang, L. Analysis of Elevator Motor Fault Detection Based on Chaotic Theory. *J. Inf. Comput. Sci.* **2014**, *11*, 229–235. [CrossRef]
4. Lan, S.; Gao, Y.; Jiang, S. Computer Vision for System Protection of Elevators. *J. Phys. Conf. Ser.* **2021**, *1848*, 012156. [CrossRef]
5. Feng, S.; Chen, J.; Liang, Y.; Xu, H. Research on Camera Bracket for Elevator Traction Wheel Groove. In Proceedings of the 2021 International Conference on Electronic Information Engineering and Computer Science, EIECS 2021, Changchun, China, 23–26 September 2021; pp. 820–823.
6. Jiang, X.Y.; Huang, X.C.; Huang, J.P.; Tong, Y.F. Real-Time Intelligent Elevator Monitoring and Diagnosis: Case Studies and Solutions with Applications Using Artificial Intelligence. *Comput. Electr. Eng.* **2022**, *100*, 107965. [CrossRef]
7. Xiaojuan, X.; Ningxiang, Y.; Jianxun, C.; Xiaoming, L. Wear Recognition Method for Traction Wheel Groove of Elevator Based on Image Processing. *China Saf. Sci. J.* **2019**, *29*, 122. [CrossRef]
8. Jia, M.; Gao, X.; Li, H.; Pang, H. Elevator Running Fault Monitoring Method Based on Vibration Signal. *Shock Vib.* **2021**, *2021*, 4547030. [CrossRef]
9. You, L.; Hu, J.; Fang, F.; Duan, L. Fault Diagnosis System of Rotating Machinery Vibration Signal. *Procedia Eng.* **2011**, *15*, 671–675. [CrossRef]
10. Esteban, E.; Salgado, O.; Iturrospe, A.; Isasa, I. Design Methodology of a Reduced-Scale Test Bench for Fault Detection and Diagnosis. *Mechatronics* **2017**, *47*, 14–23. [CrossRef]
11. Nandi, A.; Ahmed, H. *Condition Monitoring with Vibration Signals*; Wiley: Hoboken, NJ, USA, 2019; ISBN 9781119544623.
12. Wu, K.; Chu, N.; Wu, D.; Antoni, J. The Enkurgram: A Characteristic Frequency Extraction Method for Fluid Machinery Based on Multi-Band Demodulation Strategy. *Mech. Syst. Signal Process.* **2021**, *155*, 107564. [CrossRef]
13. Xiang, L.; Tang, G.; Hu, A. Vibration Signal's Time-Frequency Analysis and Comparison for a Rotating Machinery. *J. Vib. Shock* **2010**, *29*, 42–45.
14. Wang, Z.; Yang, J.; Li, H.; Zhen, D.; Xu, Y.; Gu, F. Fault Identification of Broken Rotor Bars in Induction Motors Using an Improved Cyclic Modulation Spectral Analysis. *Energies* **2019**, *12*, 3279. [CrossRef]
15. Jin, Y.; Liu, X.X.; Liu, W.P. Design of Hydraulic Fault Diagnosis System Based on Labview. *Adv. Mater. Res.* **2012**, *457–458*, 257–260. [CrossRef]
16. Feldman, M. Hilbert Transform in Vibration Analysis. *Mech. Syst. Signal Process.* **2011**, *25*, 735–802. [CrossRef]
17. Lei, Y.; Lin, J.; He, Z.; Zuo, M.J. A Review on Empirical Mode Decomposition in Fault Diagnosis of Rotating Machinery. *Mech. Syst. Signal Process.* **2013**, *35*, 108–126. [CrossRef]
18. Wang, Y.; Xiang, J.; Markert, R.; Liang, M. Spectral Kurtosis for Fault Detection, Diagnosis and Prognostics of Rotating Machines: A Review with Applications. *Mech. Syst. Signal Process.* **2016**, *66–67*, 679–698. [CrossRef]
19. Feng, Z.; Chu, F.; Zuo, M.J. Time-Frequency Analysis of Time-Varying Modulated Signals Based on Improved Energy Separation by Iterative Generalized Demodulation. *J. Sound Vib.* **2011**, *330*, 1225–1243. [CrossRef]
20. Feng, Z.; Chen, X.; Liang, M. Iterative Generalized Synchrosqueezing Transform for Fault Diagnosis of Wind Turbine Planetary Gearbox under Nonstationary Conditions. *Mech. Syst. Signal Process.* **2015**, *52–53*, 360–375. [CrossRef]

21. Chen, X.; Feng, Z. Iterative Generalized Time–Frequency Reassignment for Planetary Gearbox Fault Diagnosis under Nonstationary Conditions. *Mech. Syst. Signal Process.* **2016**, *80*, 429–444. [CrossRef]

22. Song, Y.; Liu, J.; Wu, D.; Zhang, L. The MFBD: A Novel Weak Features Extraction Method for Rotating Machinery. *J. Braz. Soc. Mech. Sci. Eng.* **2021**, *43*, 547. [CrossRef]

23. Antoni, J.; Xin, G.; Hamzaoui, N. Fast Computation of the Spectral Correlation. *Mech. Syst. Signal Process.* **2017**, *92*, 248–277. [CrossRef]

24. Antoni, J. Cyclostationarity by Examples. *Mech. Syst. Signal Process.* **2009**, *23*, 987–1036. [CrossRef]

25. Feng, Z.; Chen, X.; Liang, M. Joint Envelope and Frequency Order Spectrum Analysis Based on Iterative Generalized Demodulation for Planetary Gearbox Fault Diagnosis under Nonstationary Conditions. *Mech. Syst. Signal Process.* **2016**, *76–77*, 242–264. [CrossRef]

26. Feng, Z.; Chen, X. Adaptive Iterative Generalized Demodulation for Nonstationary Complex Signal Analysis: Principle and Application in Rotating Machinery Fault Diagnosis. *Mech. Syst. Signal Process.* **2018**, *110*, 1–27. [CrossRef]

27. Wang, Z.; Yang, J.; Li, H.; Zhen, D.; Gu, F.; Ball, A. Improved Cyclostationary Analysis Method Based on TKEO and Its Application on the Faults Diagnosis of Induction Motors. *ISA Trans.* **2022**, *128*, 513–530. [CrossRef] [PubMed]

28. Song, Y.; Liu, J.; Chu, N.; Wu, P.; Wu, D. A Novel Demodulation Method for Rotating Machinery Based on Time-Frequency Analysis and Principal Component Analysis. *J. Sound Vib.* **2019**, *442*, 645–656. [CrossRef]

29. Song, Y.; Liu, Z.; Hou, R.; Gao, H.; Huang, B.; Wu, D.; Liu, J. Research on Electromagnetic and Vibration Characteristics of Dynamic Eccentric PMSM Based on Signal Demodulation. *J. Sound Vib.* **2022**, *541*, 117320. [CrossRef]

30. Maraini, D.; Nataraj, C. Freight Car Roller Bearing Fault Detection Using Artificial Neural Networks and Support Vector Machines. *Mech. Mach. Sci.* **2015**, *23*, 663–672.

31. Singh, S.; Vishwakarma, M. A Review of Vibration Analysis Techniques for Rotating Machines. *Int. J. Eng. Res.* **2015**, *V4*, 757–761. [CrossRef]

sensors

Article

Efficient Object Detection Based on Masking Semantic Segmentation Region for Lightweight Embedded Processors

Heuijee Yun and Daejin Park *

School of Electronic and Electrical Engineering, Kyungpook National University, Daegu 41566, Republic of Korea
* Correspondence: boltanut@knu.ac.kr; Tel.: +82-53-950-5548

Abstract: Because of the development of image processing using cameras and the subsequent development of artificial intelligence technology, various fields have begun to develop. However, it is difficult to implement an image processing algorithm that requires a lot of calculations on a light board. This paper proposes a method using real-time deep learning object recognition algorithms in lightweight embedded boards. We have developed an algorithm suitable for lightweight embedded boards by appropriately using two deep neural network architectures. The first architecture requires small computational volumes, although it provides low accuracy. The second architecture uses large computational volumes and provides high accuracy. The area is determined using the first architecture, which processes semantic segmentation with relatively little computation. After masking the area using the more accurate deep learning architecture, object detection is implemented with improved accuracy, as the image is filtered by segmentation and the cases that have not been recognized by various variables, such as differentiation from the background, are excluded. OpenCV (Open source Computer Vision) is used to process input images in Python, and images are processed using an efficient neural network (ENet) and You Only Look Once (YOLO). By running this algorithm, the average error can be reduced by approximately 2.4 times, allowing for more accurate object detection. In addition, object recognition can be performed in real time for lightweight embedded boards, as a rate of about 4 FPS (frames per second) is achieved.

Keywords: autonomous driving; object detection; OpenCV; ENet; YOLO; deep learning

Citation: Yun, H.; Park, D. Efficient Object Detection Based on Masking Semantic Segmentation Region for Lightweight Embedded Processors. *Sensors* **2022**, *22*, 8890. https://dx.doi.org/10.3390/s22228890

Academic Editors: Yong Liu and Xingxing Zuo

Received: 14 October 2022
Accepted: 12 November 2022
Published: 17 November 2022

Publisher's Note: MDPI stays neutral with regard to jurisdictional claims in published maps and institutional affiliations.

1. Introduction

Currently, with the advancement of artificial intelligence technology, industries in various fields, ranging from automobiles to the Internet of Things (IoT), are developing. In these industries, artificial intelligence calculates the input of multiple datasets and converts it into the required output data [1,2]. Various types of sensors are used to receive data, among which camera sensors and methods for processing visual information input are active fields of research [3,4]. Object recognition using visual data as learning data for deep learning is used in various methods and has been researched in a variety of fields [5]. However, these data are difficult to process in real time using a processor that has small amount of memory because of the large amount of image data. In addition, to implement artificial intelligence in daily life a lightweight embedded board must be used. However, lightweight embedded boards are not suitable for large computation loads, as they have small memory and power.

The weight reduction of the object recognition algorithm using a camera sensor has always been an important task to be solved, and research is currently being conducted in various ways [6]. Various methods of processing images have been developed for effective implement of algorithms [7]. However, as with all algorithms, there is a trade-off relationship between accuracy, speed, cost, and amount of computation. The ASM framework has been studied as an effective method for mining most unlabeled or partially labeled data to enhance object detection. The ASM framework can be used to build effective

CNN detectors that require fewer labeled training instances while achieving promising results [8].

This paper introduces an object recognition algorithm based on deep learning to accurately recognize objects in real time. YOLO (You Look Only Once), a deep learning-based object recognition architecture, is currently the most well-known and efficient object recognition algorithm. However, it is too heavy an architecture to use in real-time on a lightweight embedded board. Therefore, the ROI (Region of Interest) is set in the input data to reduce the amount of image processing. Figure 1 shows the overall operation of the algorithm. The ROI can be set using ENet (Efficient Neural network), a semantic segmentation architecture based on deep learning. The the object of interest can be expressed in a specific color using semantic segmentation. By binarizing this expression, the remaining parts other than the recognized object are removed. Because this architecture only recognizes people, it is useful for removing objects other than people. Running YOLO using masked images as input data reduces computation and can be used on a lightweight embedded board, resulting in improved accuracy.

By dividing image processing into two steps in this way, the efficiency can be maximized, and the accuracy does not change rapidly in various environments. By setting the ROI after filtering using segments in the input image, cases that were not recognized by different variables, such as differentiation from the background, can be excluded, increasing the accuracy. Using two deep learning models allows for implementation with higher accuracy and faster execution time. When the amount of computation is reduced and the algorithm is implemented on a lightweight embedded board, its scope of use can be widened considerably.

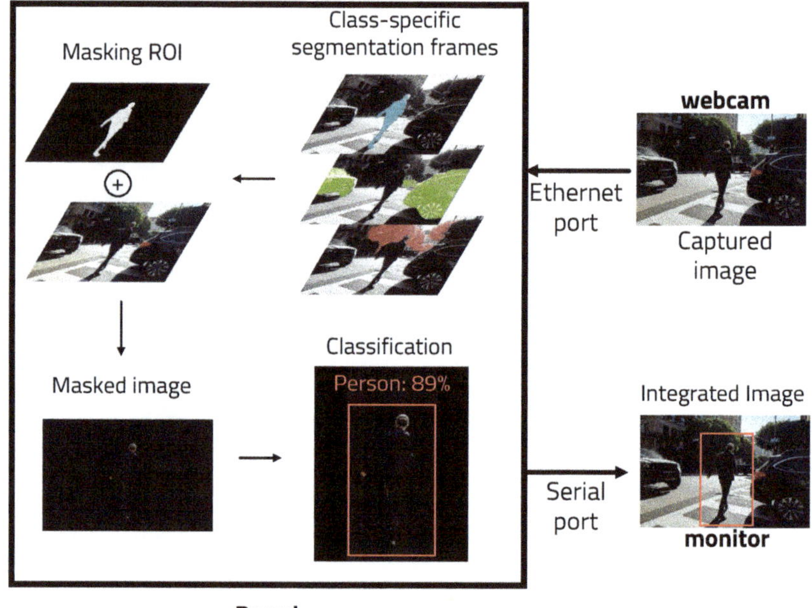

Figure 1. Overall structure.

2. Background

2.1. Preliminary Study of the Proposed Method

Deep learning is a technology that trains rules from data using artificial neural networks. An artificial neural network composed of a layer of neurons is first trained with sample data before being used to make inferences. Deep learning is trained with the help of a deep artificial neural network composed of multiple layers. As the data pass through a series of filters in a deep neural network, they can be usefully refined to handle multi-step information extraction. When analyzing images using deep learning, images are first classified. After dividing the image into defective and regular components, they are sorted and assigned by class. The image is then subdivided and each pixel is assigned to a class before image processing is performed. A significant amount of computation is required to realize the repeatability of deep learning algorithms, added complexity as the number of neural network layers increases, and implementation of the data required for training.

R-CNN (Regions with Convolutional Neuron Networks features) [9], an existing image processing system, creates a bounding box on an object by a method called region proposal, then applies a classifier to the box to classify it. After classification, it proceeds through a complex process of adjusting the boxes and removing duplicate detections. The image is then post-processed to estimate the detection probability of the boxes. Because these processes need to be trained and optimized independently, the overhead is large, requiring a significant amount of processing time.

YOLO (You Look Only Once) [10] is a real-time image detection architecture based on deep learning. Unlike RCNN, YOLO processes images in a single regression without requiring multiple steps. Using one pipeline, it can detect a target object by looking at the image once. It finds the coordinate position and probability of the bounding box in the image pixel. In addition, because the entire image is viewed and processed, it does not recognize background noise as an object, and as such the background error is small. Fast YOLO, which consists of a total of 24 convolution layers and two fully connected layers, has nine convolutional layers. YOLO convolution layers can be trained with datasets. YOLO uses a framework called darknet to enable training and inference. Figure 2a shows a box drawn around the recognized object, with its class and probability shown in the kernel.

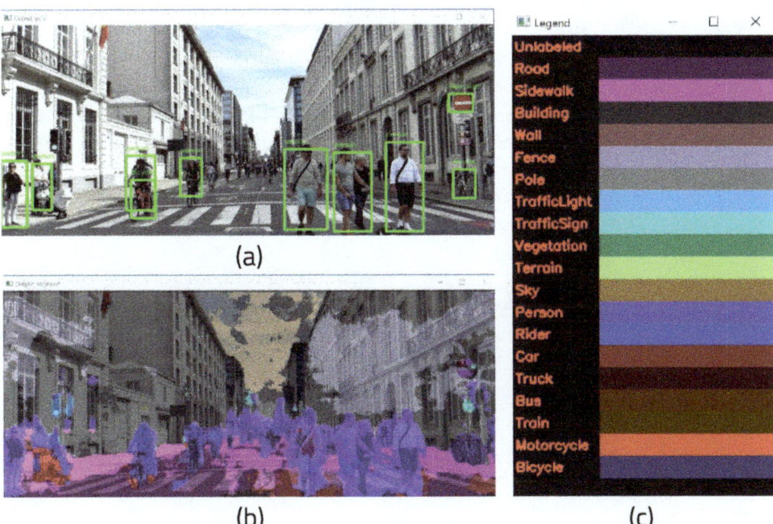

Figure 2. (**a**) Result of YOLO object detection, (**b**) result of semantic segmentation, and (**c**) classes and labels of semantic segmentation.

Semantic segmentation [11] classifies all pixels in an image into a designated class to recognize objects. It is currently used in various fields, including self-driving cars and medical image analysis. The difference between SS and object recognition algorithms such as YOLO is that the former determine which class the pixels themselves belong to. As a result, the number of people or objects in the image cannot be counted; only the types of recognized objects can be identified.

There are numerous algorithm models for implementing semantic segmentation, of which the fully convolutional network (FCN) approach [12] is the most well-known. Because an FCN only consists of convolutional layers, it is not necessary to fix the size of the input image. In addition, because the filter is learnable and stems from a single deep learning model, an end-to-end model can be used. Because it processes the entire image at once, it can be processed quickly. However, it has the disadvantage of lower the accuracy when the resolution changes during the processing process. ENet (Efficient Neural Network) [13] is a deep learning-based semantic segmentation structure that uses an algorithm model called ResNet [14]. Although the existing method requires a large amount of computation, consumes a lot of power, and has a slow processing speed when classifying object classes in units of pixels, ENet has developed a new deep neural network structure that can be performed on an embedded board. ENet is made up of an encoder and a decoder along with thirteen convolution filters. A sampling operation is used to solve the problem of resolution loss and increase the accuracy. As shown in Figure 2b,c, objects within the image can be classified using the color assigned to each class.

2.2. Concepts and Definitions of the Proposed Method

The method devised in this paper involves the design of a more lightweight object recognition algorithm using the two deep learning models described above. A filter is created for object recognition in the image. This filter recognizes the pixels an object occupies in the image, and masks only the object. The more complex the background excluding the object in the image, the lower the accuracy and the greater the amount of computation. Therefore, it is possible to reduce the amount of computation by recognizing the filtered image as an object.

This study conducted further experiments based on previous studies [15]. In the mentioned study, the authors used two lightweight embedded boards and measured for memory and time as well as for various elements such as background complexity and power consumption. They were able to identify efficiency and accuracy by experimenting with various versions of YOLO that recognize objects.

There have been other studies using YOLO by setting ROI as segmentation. In [16], the authors used YOLO to set ROI by segmentation and then executed object recognition with CNN. Similar to the present study, they implemented it in two stages. The structure was used for interpreting sign language with deep learning. After extracting hand parts using YOLO, sign language was interpreted by a CNN. When learning by setting ROI, it is apparent that large amounts of data are not needed and that the speed and success rate are improved.

3. Implementation

We propose a method to execute the object recognition algorithm in real time on a lightweight embedded board by writing a lightweight and divisional-sized algorithm. Two deep learning-based systems were used to reduce the amount of computation and increase the accuracy and overall structure, as can be seen in Figure 3.

The reason why a model that integrated segmentation and object detection was not used from the beginning was to divide the code. There is benefit in separating segmentation and object detection. First, ENet and pre-image postprocessing can be operated on a very small FPGA board, and YOLO can be run on a better performance FPGA. Combining segmentation and detection into one model requires an embedded board that performs much better than is necessary when running each individually. Two boards with relatively poor performance are much more cost-effective than one high-performance board. In addition, the communication time between the two FPGA boards is very short (less than 0.1 s), and the communication cost is insignificant because it uses less energy, which is more effective.

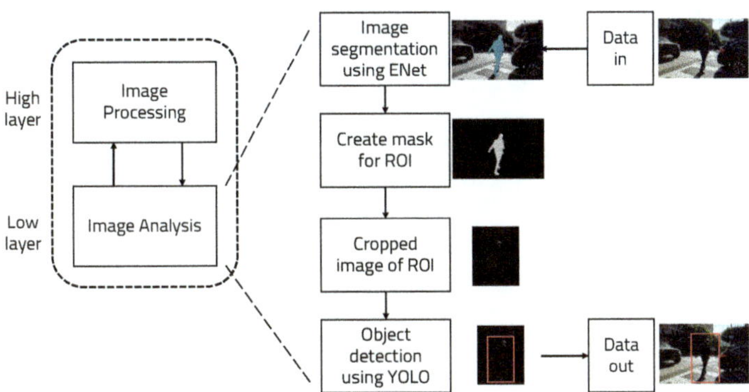

Figure 3. Overall algorithm used for real-time object detection.

3.1. Semantic Segmentation Object Detection

Figure 4 shows the diagram of the semantic segmentation algorithm using ENet. The video data received by the webcam are captured as frame units. One frame is captured every 0.03 s on average. The frame is masked using the trained weights. Using trained weights, a human in an image can be detected and segmented. This allows the ROI to be selected for object recognition in the image. We used ENet because it is easily trained to detect specific targets and has higher accuracy. When training ENet, the UP-S31 dataset from the Leeds Sports Pose dataset with the MPII Human Pose Dataset was used to recognize humans [17].

Currently, with the rapid development of deep learning and computer vision fields, many highly diverse models are being developed, such as DUC-HDC (Dense Upsampling Convolution and Hybrid Dilated Convolution) and DCNAS (Densely Connected Neural Architecture Search). However, it was difficult to find a model that could be be executed simultaneously in different in operating systems, taking into account the differences in the board and the host PC and their corresponding execution methods. Pytorch, Mask-RCNN, and ENet, were judged to be the most stable and widely known for the two operating systems. Thus, these were the main models tested, and the study was conducted according to the results.

We tested with pytorch, mask RCNN, and ENet. For a single frame, ENet took 1.75 s, Pytorch took 4.45 s, and Mask RCNN took 1.22 s. We tested these models on the same PC based on Windows 10. The input image was 640×959 in size and was a jpeg file. In order to check whether segmentation was performed well, an image containing only one person was tested. The accuracy was 0.89 for ENet, 0.92 for Pytorch, and 0.84 for Mask RCNN. Figure 5 shows the results with each algorithm. Eventually, ENet was chosen, as it was most efficient in terms of time and accuracy. Another model, IC-Net, [18] was tested as well. The time and accuracy of IC-Net were calculated using the same Cityscapes dataset. The average FPS (frames per second) rate is the time taken per frame, and can be calculated by dividing the total time taken to process the image by one. The average FPS of IC-Net on the host PC was 1.196, and the average FPS rate on the LS1028 board was 0.1402. Accuracy can be calculated by dividing the total number of people recognized after image processing by the number of people actually in the image. The accuracy was calculated as 0.73. Figure 6 shows the results with the different frameworks segmentation algorithm when using weights trained with the same Cityscape dataset; (a,b) show the results for ENet, and (c,d) show the results for IC-Net. Although the accuracy of the two models is similar, there are many differences in terms of FPS. Although the accuracy of segmentation is low, because image processing is performed again after ROI setting it is not necessary to sacrifice power or memory consumption for higher FPS and accuracy.

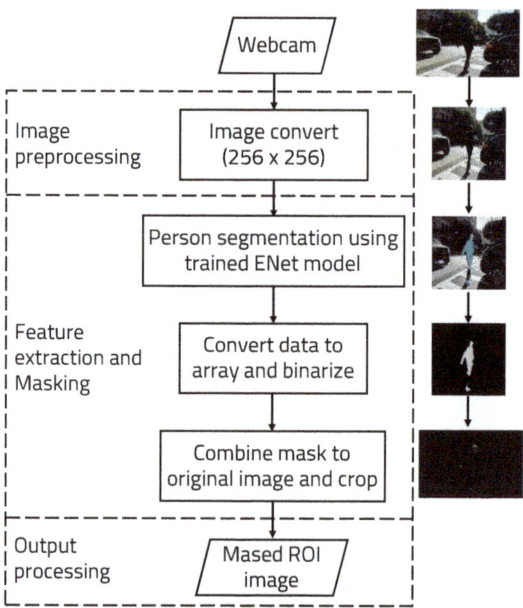

Figure 4. Algorithm used for semantic segmentation.

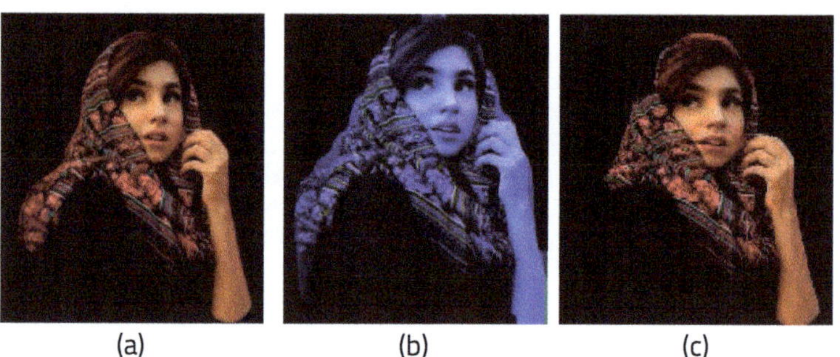

Figure 5. Results of segmentation algorithms: (**a**) Pytorch, (**b**) Mask RCNN, (**c**) ENet.

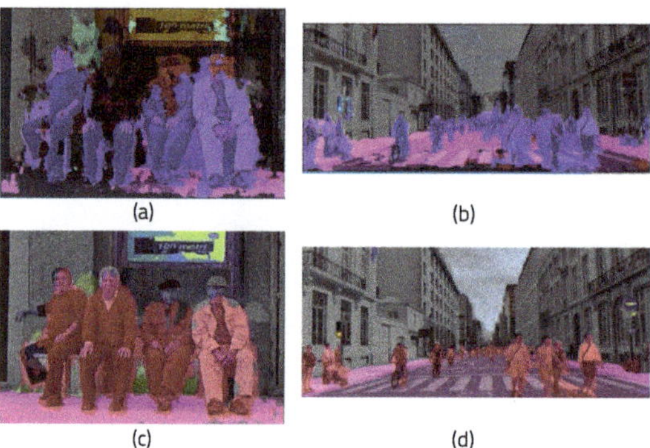

Figure 6. Results for IC-Net and ENet: (**a**,**b**) results for ENet; (**c**,**d**) results for IC-Net.

Algorithm 1 shows the semantic segmentation algorithm in more detail in the form of pseudo-code. This algorithm makes it easy to process the image received from the webcam by dividing it into frames utilizing OpenCV (Open-Source Computer Vision) functions.

A trained model used in ENet is loaded using Keras and Tensorflow [19,20]. Because the weights trained in ENet process images with a size of 256 × 256, the initial width of the image should be 256. To obtain the memory usage within a Python program, the psutil Python package was used to create a function that outputs the memory in MB.

Algorithm 1: Pseudo-code of semantic segmentation algorithm

```
 1  cap = VideoCapture
 2  Model = load model("model path")
 3  img_wh = 256
 4  Function Memory_usage(debug):
 5      rss = memory_info() / 2 ** 20
 6      print("Memory usage(MB)", rss)
 7  Function Seg(img):
 8      while True do
 9          ret, orig_img = cap.read()
10          img = resize(orig_img, (img_wh, img_wh))
11          img_tensor = expand_dims(img, 0)
12          raw_output = model.predict(img_tensor)
13          seg_labels = [array of object detection label results]
14          seg_img = max(seg_labels, axis = 2)
15          mask = zero array of 256 × 256
16          area = 0
17          area_back = 0
18          sum_back = 0
19          v_sum = 0
20          foreach i ∈ (0, 255) do
21              foreach j ∈ (0, 255) do
22                  if seg_img[i][j] != 0 then
23                      mask[i][j] = 1
24                      area += 1
25                  else
26                      sum_back += img[i][j]
27                      area_back += 1
28          mean = sum_back/area_back
29          foreach i ∈ (0, 255) do
30              foreach j ∈ (0, 255) do
31                  if seg_img[i][j] = 0 then
32                      vsum = vsum + (img[i][j] − mean) ** 2
33          variance = vsum/area_back
34          img_dis = array(img)
35          output = bitwise_and(img_dis, mask)
36          memory_usage()
```

We created a function that receives the frame image from the webcam as input and performs segmentation. It first receives the original frame size, then changes the webcam frame size (which was 640 × 480) to 256 × 256, the width of the initialized image. The batch dimension can then be set using the NumPY function. ENet loads a model trained to recognize only humans and runs it. Because the segmentation result is colored based on the recognized person, only the colored part needs to be extracted for masking. First, the labeled part of the result is converted into an array and combined with the image array. Before masking, the mask array and variables are initialized to calculate the amount of computation based on the complexity of the background. If the array of the segmented result image for masking is not 0, the mask array is set to 1 in order to binarize it. Figure 7a shows the original image, and Figure 7b shows the binarized mask showing only the part of the image recognized as a person.

(a)   (b)

Figure 7. Results of masking: (**a**) original frame and (**b**) masked image.

In addition, the area variable is set to 1 whenever a pixel occupied by an object is found in order to determine how much space the recognized object takes up in the background. This helps to measure how image background complexity affects the performance of the algorithm. If the data of the segmented pixel is zero, it means that the pixel has no data on the segmented object. Therefore, it can be judged as the background of the image. To calculate the complexity of the background, the pixel value of the original image is added to obtain the background variance, then 1 is added to area_back. To determine the variance of the RGB values of the background, first, the average value of the background is obtained, then the sum of the deviations is computed. The variance can be calculated by dividing the sum of the deviations by the total number of elements in the entire background. To create a masked image, the original image is converted into an array, then the mask and the original image are combined using the bitwise and operation functions in OpenCV. Figure 8 shows the masking result by combining the original image and the mask. The memory usage function that was previously created is called to determine the computational cost of this process.

Figure 8. Results of ROI masked image.

3.2. Object Detection with YOLO

Figure 9 shows the structure of the YOLO algorithm, which recognizes an object in an image that was previously masked using ROI with segmentation. The YOLO model recognizes the masked image, and the box and probability of the recognized object are displayed.

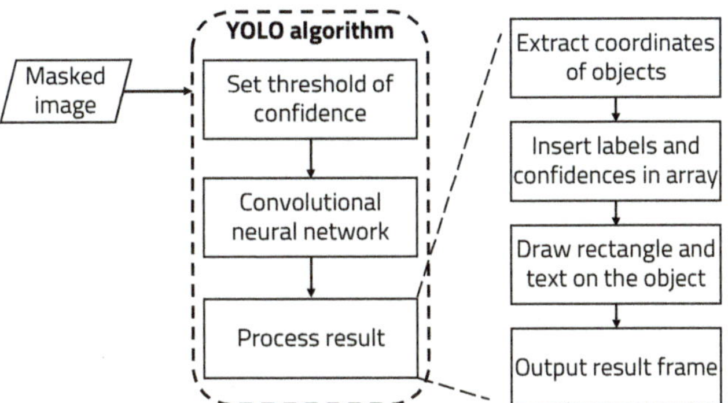

Figure 9. YOLO algorithm for object detection.

As a first stage target, detection algorithms are implemented in a way that is lightweight and can be operated quickly and accurately; two algorithms, YOLO, and SSD, were tested [21] while running region proposal and classification simultaneously. It is not necessary to use a neural network called RPN (Region Proposal Network) to generate candidate targets. These algorithms predict target locations and classes directly over the network. As it is a method for solving the classification and localization problems at the same time, it can be simulated in high FPS. An efficient deep neural network model called MobileNets was tested as well [22]. MobileNet is an efficient convolutional neural network for low power devices. MobileNet utilizes depth-wise separable convolution to make the model lightweight. Two parameters were used to optimally fit MobileNet in memory-constrained environments. These two parameters adjust the balance between latency and accuracy. For better variety in the experiment, Faster R-CNN [23], a two stage detector, was tested as well. In a two-stage detector, regional proposals and classification are performed sequentially. The R-CNN algorithm has a limitation in that it is slower than YOLO. Fast R-CNN greatly reduces iterative CNN computations; however, the region proposal algorithm becomes the bottleneck. Faster R-CNN uses an RPN (Region Proposal Network) in the region proposal process to make the existing Fast R-CNN faster. Figure 10 shows the results of three algorithms. They were tested on the same PC based on Windows 10, and the weights were all trained using the Pascal VOC dataset. For simple detection, the input image was converted to 512×512. SDD with Pytorch took 7.6 s, YOLO took 4.24 s, MobileNet took 0.09 s, and Faster R-CNN took 7.3 s. The run time of the MobileNet model was overwhelmingly short, while the accuracy between YOLO and Faster R-CNN was quite similar. However, for the second stage of object detection, it is necessary to operate with high accuracy. As can be seen from the results, YOLO is the most accurate algorithm, and is fast as well. Eventually, YOLO was chosen for the second stage of object detection for subsequent experiments.

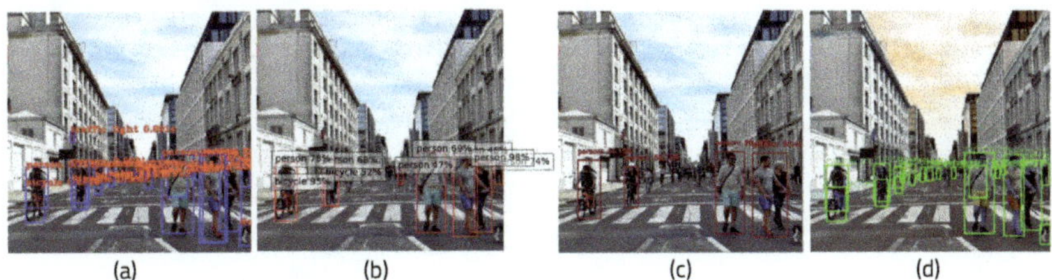

Figure 10. Results for stage one target detection algorithms: (**a**) YOLO, (**b**) SSD, (**c**) MobileNet, (**d**) Faster R-CNN.

Algorithm 2 shows the YOLO object recognition algorithm in pseudo-code. YOLO is an algorithm developed on a C++-based framework called darknet in Linux. However, in order to use Tensorflow and various Python packages, darkflow [24], which ports YOLO to other Tensorflow-based platforms, was used. It is convenient because the number of codes is reduced and multiple functions can be used. In addition, it is faster than darknet. Using darkflow in Tiny-YOLO weight executes in 1.3 fps, while darknet takes 0.3 fps [25]. To run darkflow, the model's path and weights are first set to be executed in YOLO. We used the YOLOv3 and Tiny-YOLOv3 weights, as they are stabilized for darkflow. YOLO is being developed in various ways, such as YOLOv5 and PP-YOLO. However, there were problems in the process of building and running it on the board, which are currently being resolved. We plan to present a more advanced architecture in the future. Even if the confidence threshold is set low, the accuracy hardly changes when setting the ROI. The TFNet class object was initialized with the previously set values to build the model. Random numbers were generated to determine the color to be used when displaying the recognized object.

By receiving an image with ROI masked by segmentation as input, the image can be analyzed with the previously built model using TFNet. The current time is recorded in a variable called start_time to calculate the processing time and fps. This initializes obj_count, a variable that counts the number of people to calculate accuracy. In order to show the result of the analyzed image, the coordinates of the upper part of the recognized object are stored as the tl variable, and the coordinates of the lower part of the object are stored as the br variable. The class and probability of the recognized object are stored as the text variable, so called to make it easily printed into the image.

Algorithm 2: Pseudo-code for YOLO object detection.

```
1  Options = {
2       model : "model path"
3       load : "weight path"
4       threshold : 0.3
5  }

6  tfnet = TFNet(Options)
7  color = (255 * rand(3) in range (10))

8  Function YOLO(masked):
9      while True do
10         start_time = time()
11         results = tfnet.return_predict(masked)
12         obj_count = 0
13         foreach color, result ∈ (color, results) do
14             tl = (result['topleft']['x'], result['topleft']['y'])
15             br = (result['bottomright']['x'], result['bottomright']['y'])
16             label = result['label']
17             confidence = result['confidence']
18             text = ': '.(label, confidence * 100)
19             output = draw_rectangle(masked, tl, br, color, 3)
20             output = put_text(output, text, tl, font, 0.25, 1)
21             obj_count += 1

22         show('frame', output)
23         print('people', obj_count)
24         print('FPS', 1/(time() - start_time))

25         memory_usage()
```

A box is drawn on the recognized object using the values stored in the variables tl and bl; then, the class and probability of the object are stated in the upper part of the box. Each time a box is drawn, one object is added to obj_count to count the number of recognized objects. In order to display the result again in real-time, the output is displayed using the

OpenCV function. Figure 11 shows the resulting image recognized as only humans using YOLO in the ROI set by semantic segmentation. The number of people recognized and the fps can be calculated and printed. Using the memory usage calculation function written in the semantic segmentation algorithm, the amount of computation when processing images using YOLO can be calculated.

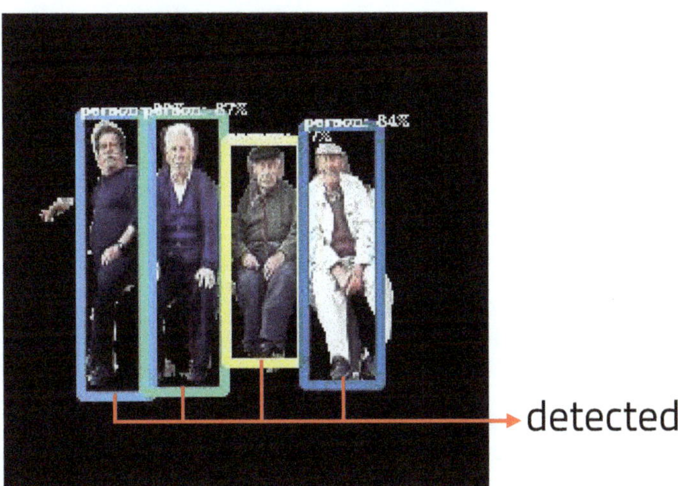

Figure 11. Results of YOLO object detection.

4. Experimental Setup and Evaluation

This section describes the image classification neural network target program and evaluates the proposed fault detector performance with random test images and fault injections.

4.1. Experimental Environment

Experiments with object recognition algorithms based on semantic segmentation and YOLO were conducted on the LS1028a [26] and LX2160a [27] boards as well as a Windows 10 PC. The boards were configured as shown in Figure 12. In order to determine whether the algorithms could be run on multiple operating systems, they were run on a Linux-based board and a Windows-based board for the experiments. Because these two operating systems have different methods of executing programs, memory, and performance on each board, various results were obtained. The LS1028a and LS2160a are processors made by the NXP company. The LS1028ardb is equipped with two 64-bit ARM Cortex-A72 processors with a maximum operating speed of 1.3 GHz per core, while the LX2160a is equipped with sixteen 64-bit ARMv8 Cortex-A72 processors, and the maximum operating speed per core is 2.2 GHz. They can both manage input/output and communication through an ethernet port, USB port and serial port. Python code was used for measuring execution time, memory usage, FPS, and number of detected people. To measure power consumption, Open Hardware Monitor was for host PC (Windows) and the Powertop tool was used for the Linux FPGA board.

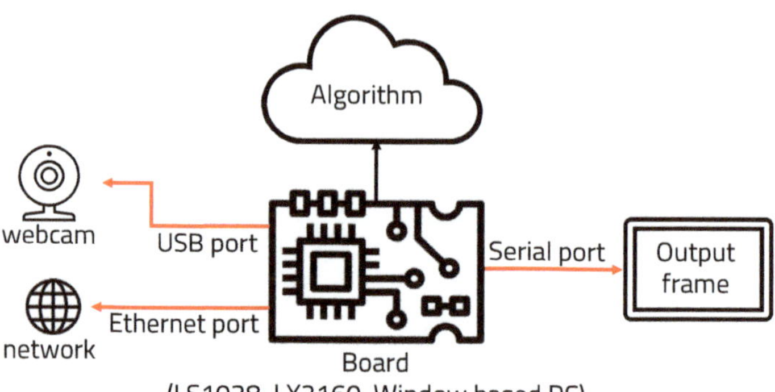

Figure 12. Structure of boards and algorithms.

4.2. Execution Time and Accuracy

4.2.1. Methods for Measuring the Accuracy of Deep Learning Models

There are several methods for measuring the accuracy of a deep learning model; here, we used a confusion matrix [28]. The confusion matrix can be divided into four states: TP, when the model corrects the correct answer; TN, when the model predicts the correct answer incorrectly; FP, when the model incorrectly predicts the correct answer as an incorrect answer; and FN, when the model incorrectly predicts the incorrect answer as an incorrect answer. Equation (1) is the formula for calculating accuracy:

$$\text{Accuracy} = \frac{\text{TP} + \text{TN}}{\text{TP} + \text{TN} + \text{FP} + \text{FN}}. \tag{1}$$

In addition, precision, recall, and F1-score were calculated for various performance evaluation certificates. Precision was calculated using Equation (2), recall was calculated with Equation (3), and F1-score was calculated with Equation (4). The precision, recall, and the following formulas were used to calculate F1-score. We calculated TP as the number of people detected by YOLO, FP as the number of detected objects that were not human, and FN as the number of people that were not detected.

$$\text{Precision} = \frac{\text{TP}}{\text{TP} + \text{FP}}, \tag{2}$$

and

$$\text{Recall} = \frac{\text{TP}}{\text{TP} + \text{FN}}, \tag{3}$$

and

$$\text{F1-score} = \frac{2}{\frac{1}{\text{precision}} + \frac{1}{\text{recall}}}. \tag{4}$$

4.2.2. Time and Measurement Accuracy Results of the Proposed Algorithm

Figure 13 shows the results of measurements of time, fps, and the number of people detected when running only YOLO and when running YOLO and ENet on a Windows 10-based PC. Because there is a limit when measuring places with a large change in the number of people in real time, this was experimented with by showing pictures including different people on a webcam. Changes were measured in terms of FPS, power consumption, and memory usage while turning seventeen pictures containing different numbers of people into a slide show with a real-time webcam, which is demonstrated in Figure 14. When YOLO is used alone, fps and processing time values fluctuate according to the changing number of people. This means that if the floating population is high, operating YOLO alone is not be stable. The dispersion of fps can be derived by calculating the average fps and subtracting each fps value from the average, then adding all differences from the mean and dividing by the number of fps values. Here, the average fps is 3.64 and the dispersion of fps is 0.166. However, it can be seen that the fps and processing time are stable regardless of the changing number of people when integrating ENet and YOLO, as the average fps is 3.93 and the dispersion of fps is 0.0051.

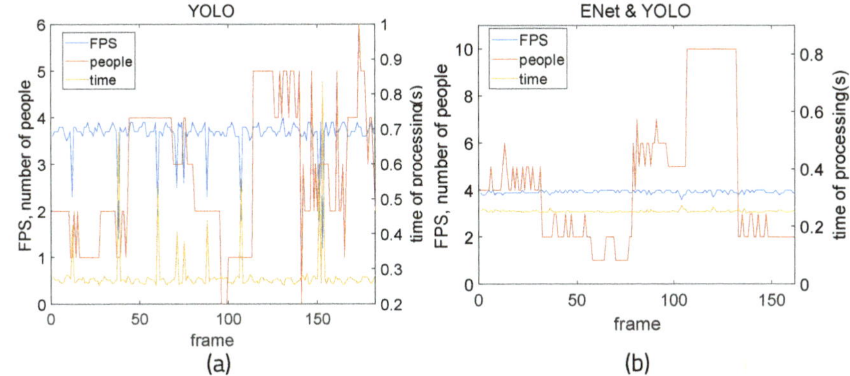

Figure 13. Time measurement: (**a**) YOLO and (**b**) YOLO executed with ENet.

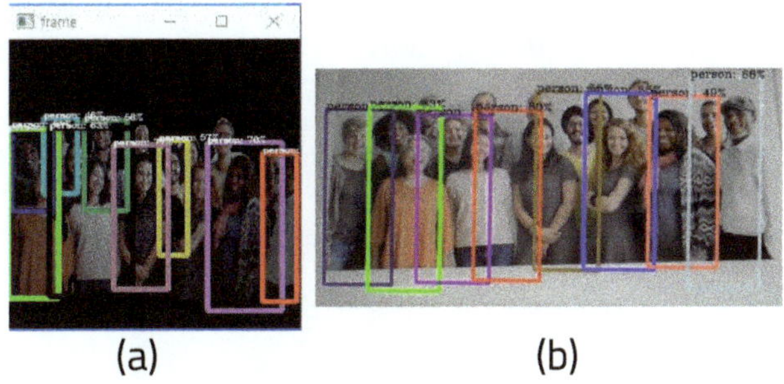

Figure 14. Result of detection of numerous people: (**a**) YOLO executed with ENet and (**b**) YOLO.

Figure 15 shows the measurement results of average time, fps, accuracy, and error when using YOLO alone and when using ENet and YOLO together running on a Windows 10-based PC. With far fewer convolutional layers than YOLO and a fast and compact encoder–decoder structure, ENet's average processing time was 0.156 s per frame, and the average fps was 6.41. When only YOLO was used, the average processing time per frame was 0.281 s and the average fps was 3.645 s. When ENet results were processed with YOLO, the average time was 0.269 s and the average fps was 3.701.

As can be seen, the method of setting ROI using ENet and processing it as YOLO input can reduce YOLO's execution burden as a result of lowering the threshold of YOLO. The total number of recognition frames was divided by the number of people in the image and the number of frames correctly recognized to determine the accuracy of the object recognition algorithms. A clear improvement in accuracy can be seen when using YOLO and ENet together. Because TP and TN accurately calculated the part where the number of people matched, the error was calculated to obtain more appropriate accuracy by comparing the recognized number of people and the actual number of people. When only YOLO was used, the average recognition error was 7.097, and when ENet and YOLO were used together, the average recognition error was 2.913. When more people were in the picture, more errors appeared when using YOLO without ENet. However, when ENet and YOLO were used together, errors were be reduced and accurate recognition was achieved.

Algorithms	Average time	Average fps
ENet	0.156s	6.41
YOLO	0.281s	3.645
ENet + YOLO	0.269s	3.701

(a)

Algorithms	Accuracy	Average error of counting	Precision	Recall	F1-score
ENet + YOLO	0.574	2.913	0.992	0.806	0.866
YOLO	0.549	7.097	0.937	0.72	0.834

(b)

Figure 15. Time and accuracy measurement: (**a**) time and FPS measurement and (**b**) accuracy measurement.

4.3. Experimental Results: Memory Consumption

Figure 16 shows the results of measuring the amount of memory used when running the algorithms on the Windows 10-based PC. Figure 16a shows the memory measurement of YOLO alone, and Figure 16b shows the memory usage of YOLO and ENet. When YOLO detects an object, it is judged as a matching object when the threshold is 0.5 or more through the IoU (Intersection Over Union) of the Bounding Box and the Correct Answer Box, which contain information about the predicted object. The higher the threshold, the more consistent the correct answer; it is important to set an appropriate threshold, because when the reference threshold is too high, the detection rate is lower [29]. It can be seen that minimal memory is used at a certain threshold. The reason for this is that when the threshold is too low, there are a lot of objects detected and consume a lot of memory. However, when the threshold is too high, no objects are detected, and the memory usage is low. When the threshold is 0.4, the average memory usage is 2.067 GB.

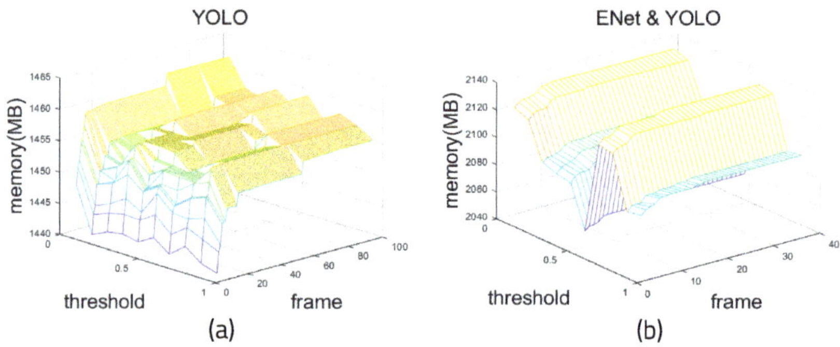

Figure 16. Relationship between memory usage and threshold: (**a**) YOLO and (**b**) ENet and YOLO.

When using ENet and YOLO together, the memory usage is obviously higher than when using YOLO alone because the models for the two architectures must be loaded separately. Despite the fact that the number of memory bytes is increased, the error is greatly decreased. In addition, this method is valid because it can be used effectively on an embedded board.

4.4. Experimental Results: Background Complexity

Figure 17 shows how the complexity of the background in the picture affects the calculation when executed on the Windows 10-based PC. The background was considered to be complicated if the variance was large by calculating the average of the RGB values and the resulting deviation and variance in the background aside from the recognized person. First, photos with complex backgrounds and people on a white background were visually selected and entered into the webcam in real time. These were then measured by dividing the complex background and the simple background into twenty photos each.

Figure 18 shows example images of background complexity. The variance of the RGB values is 0.047 in Figure 18a and 0.132 in Figure 18b. When the background is simple, the average FPS is 3.46, the average memory is 2096.937 MB, and the average variance of the RGB values is 0.0655. When the background is complex, the average FPS is 3.17, the average memory is 2133.625 MB, and the average variance of RGB is 0.177. Because the difference in the RGB average variance is large, the complexity of the background can be distinguished by the variance value. When performing semantic segmentation using ENet, it can be seen that it is not necessary to exclude the background of the photos. Because the background is simple, segmentation requires less time and less memory.

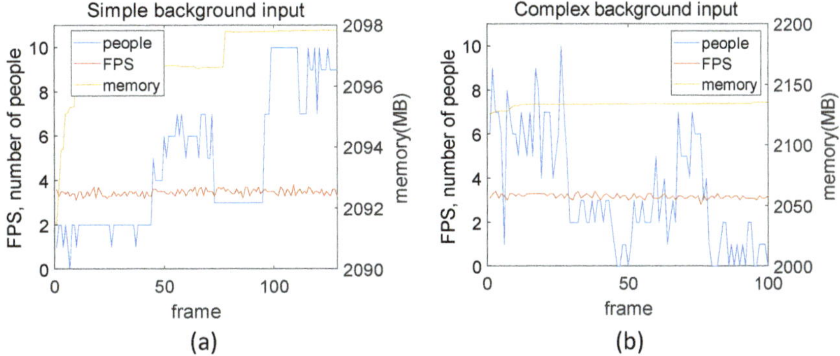

Figure 17. Time measurement result of complexity of background: (**a**) simple background and (**b**) complex background.

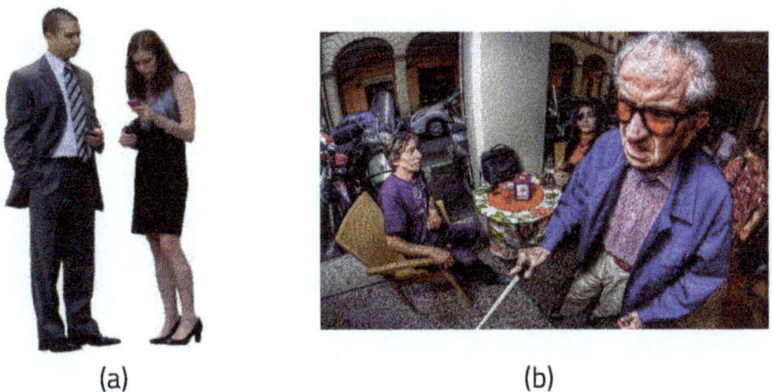

Figure 18. Example image of complexity of background: (**a**) simple background and (**b**) complex background.

4.5. Experimental Results: Performance on Boards

Figure 19 shows the measurement results when running on the LS1028a board. The Tiny-YOLO weights [30] were used because light weights should be used on the board. Tiny-YOLO is lighter than YOLO; while it has lower accuracy, it is more suitable because of its small size. When using Tiny-YOLO alone, the average FPS is 1.1 and the average memory usage is 930.08 MB. Calculating the accuracy using the previous method, the average accuracy is 0.2569 and the average error is 1.88. The precision is 0.997, recall is 0.539 and F1-score is 0.699. When using ENet and Tiny-YOLO together, the average FPS is 1.3 and the average memory usage is 1320.08 MB. The mean accuracy is 0.5866 and the mean error is 1.52. Precision is 0.968, recall is 0.711, and F1-score is 0.828. It can be seen that FPS, accuracy, and F1-score all improved when ENet was integrated.

FPS was reduced when running on a much lighter board than a PC. Using Tiny-YOLO weights, the amount of memory is reduced by almost half; however, it can be seen that there is a big difference in accuracy between the two algorithms. When more weights with low accuracy are used, the accuracy is increased by setting the ROI using ENet. For comparison, in Figure 20, the Windows-based PC was measured using Tiny-YOLO weights. When using YOLO-Tiny and ENet, the average fps is 17.57 and the average memory usage is 1290.75 MB. The mean accuracy is 0.544 and the mean error is 1.46. Precision is 0.997, recall is 0.709, and F1-score is 0.828. When using Tiny-YOLO alone, the average fps is 16.72 and the average memory usage is 642.33 MB. The mean accuracy is 0.263 and the mean error is 3.44. Precision is 0.977, recall is 0.389, and F1-score is 0.556. It can be seen that Tiny-YOLO uses less memory and has higher fps than using YOLO alone, while when ENet is integrated, the accuracy and fps are enhanced and the F1-score increases.

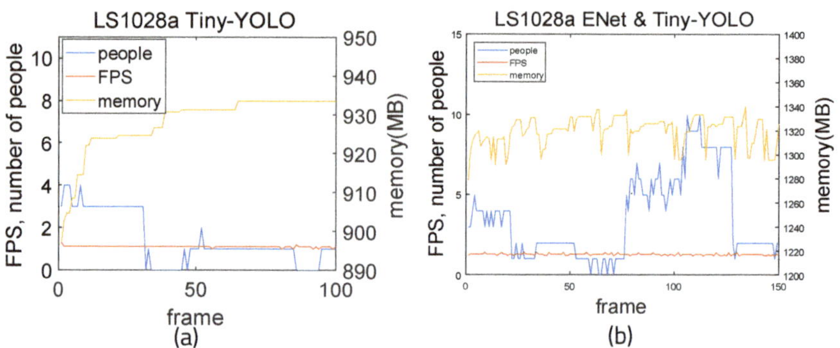

Figure 19. Measurement results for LS1028a board: (**a**) Tiny-YOLO and (**b**) ENet and Tiny-YOLO.

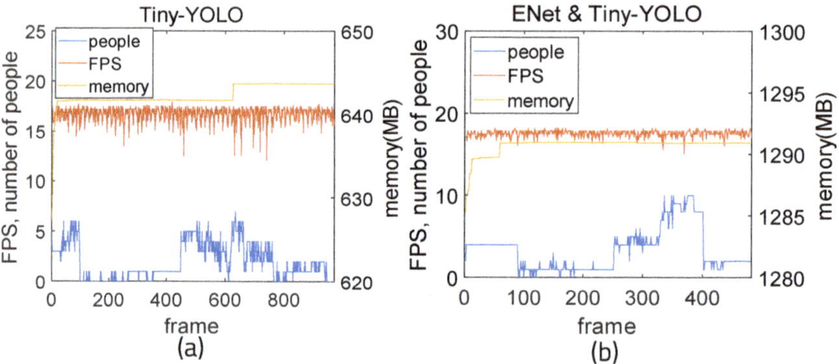

Figure 20. Time measurement results for Tiny-YOLO: (**a**) Tiny-YOLO and (**b**) ENet and Tiny-YOLO.

Figures 21 and 22 show the measurement results when running on the LX2160a board. Both YOLO weights and Tiny-YOLO weights were used, as this board has more memory than the LS1028a board. When using YOLO alone, the average FPS is 1.643 and the average memory usage is 2775.02 MB. By calculating the accuracy using the previous method, the average accuracy is 0.13 and the average error is 1.72. Precision is 0.937, recall is 0.768, and F1-score is 0.844. When using ENet and YOLO together, the average FPS is 1.664 and the average memory usage is 2855.165 MB. The mean accuracy is 0.1553 and the mean error is 1.25. Precision is 0.973, recall is 0.788, and F1-score is 0.87. When using Tiny-YOLO alone, the average FPS is 7.967 and the average memory usage is 1527.21 MB. The average accuracy is 0.178 and the average error is 2.6. Precision is 0.88, recall is 0.629, and F1-score is 0.87. When using ENet and Tiny-YOLO together, the average FPS is 8.96 and the average memory usage is 1692.72 MB. The mean accuracy is 0.157 and the mean error is 2.42.

Precision is 0.943, recall is 0.662, and F1-score is 0.777. It can be seen that when using ENet and YOLO together, fps is higher and error is lower. Furthermore, the difference in memory usage is small at 100 MB. When using ENet and Tiny-YOLO, fps values are much higher than for YOLO alone. The gap between memory usage of ENet with Tiny-YOLO and Tiny-YOLO alone is small.

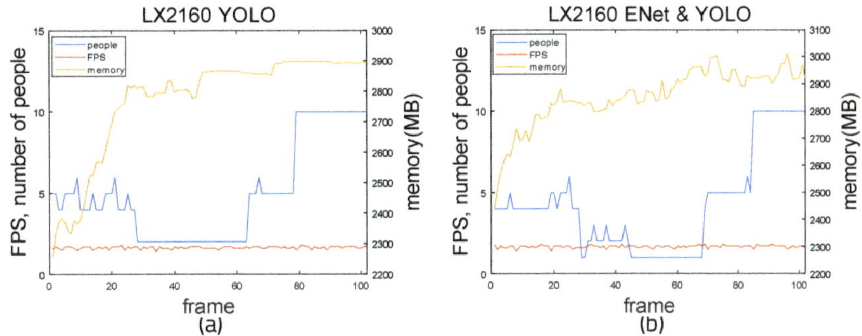

Figure 21. Time measurement result of LX2160 board: (**a**) YOLO and (**b**) ENet and YOLO.

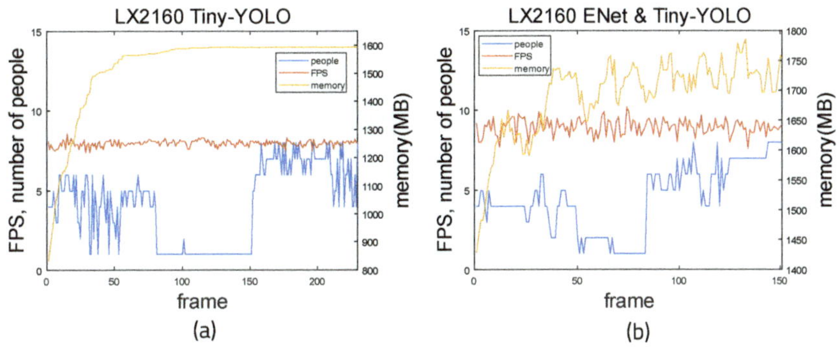

Figure 22. Time measurement result of Tiny-YOLO weight on LX2160 board: (**a**) Tiny-YOLO and (**b**) ENet and Tiny-YOLO.

4.6. Experimental Results: Power Consumption

Figure 23 illustrates the power consumption of each algorithm. They were executed on the host PC, and the threshold was fixed to 0.4. The power consumption of the architectures was determined running on an AMD Ryzen 7 3800XT 8-Core Processor with 3.90 GHz, RAM 32.0 GB, and Windows 10 Pro. The power consumption results are the average value of the total CPU power used by the architecture when only the CPU is used. The amount of power used by the architecture to process the same data was measured by putting the using picture as the input for the same period of time. It was not measured on the boards, because it was only necessary to compare the validity of the architectures' power usage. When using ENet and YOLO, the power consumption is 22.54 W, while when using YOLO alone is 32.831 W. Executing ENet and Tiny-YOLO consumed 14.14 W, while Tiny-YOLO alone consumed 26.716 W. It can be seen that using ENet with YOLO reduces power consumption by approximately 10 W, which is useful for lightweighted embedded boards, as Tiny-YOLO uses less power than YOLO.

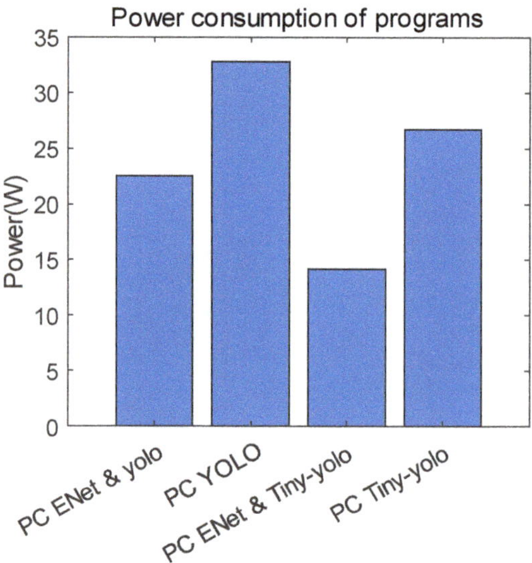

Figure 23. Power consumption results.

4.7. Experiments in Different Deep Learning Frameworks

To evaluate the effects of the architecture, it was evaluated with various deep learning frameworks. Figure 24 shows images resulting from the different frameworks. As there are two types of state-of-the-art target detection algorithms based on CNN, SSD and Faster R-CNN are assessed. SSD is a one-stage end-to-end target detection algorithm. Faster R-CNN is a two-stage target detection algorithm, in which the process is divided into two phases. A lightweighted architecture, MobileNet, is tested as well, as it can be performed efficiently on the boards. Figure 25 illustrates the time and accuracy measurement of these frameworks. The results when executing the algorithm using YOLO on the host PC and the FPGA board were similar in terms of the trends in the two environments; thus, the experiment was conducted only on the host PC. The time and FPS variance between the masked input image and the image without masking was 2.1 times faster with SSD, 1.5 times with MobileNet, and 1.4 times with Faster R-CNN. There was not much difference in the accuracy or F1-scores of the frameworks. However, the average error fell by about half in each framework. Judging from these results, it can be concluded that the proposed algorithm is efficient for object detection processing.

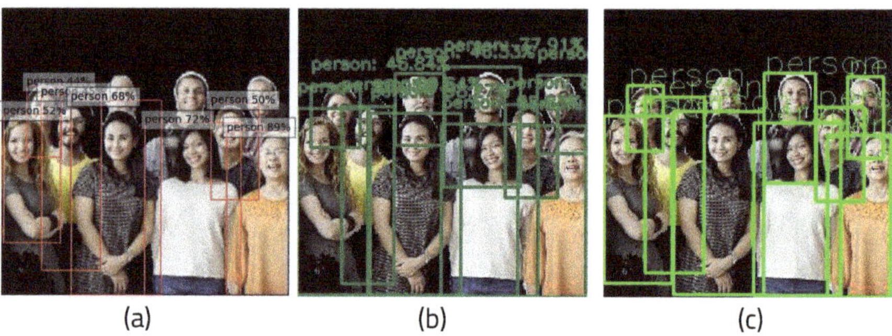

Figure 24. Results for different frameworks: (**a**) SSD, (**b**) MobileNet, and (**c**) Faster R-CNN.

Algorithms	Average time	Average fps	Accuracy	Average error	F1-score
SSD without mask	4.2s	0.23	0.53	5.3	0.83
SSD with mask	1.91s	0.52	0.56	2.16	0.85
MobileNet without mask	0.12s	8.1	0.51	4.3	0.86
MobileNet with mask	0.08s	12.6	0.56	2.83	0.87
FasterRCNN without mask	6.1s	0.16	0.65	3.67	0.88
FasterRCNN with mask	4.23s	0.23	0.67	1.66	0.89

Figure 25. Time and accuracy measurement of different frameworks.

There are several versions of YOLO, which has been developed up to version 7 as of 2022. YOLOv4, v5, v6, and v7 were tested with the proposed algorithm. YOLOv4 [31] increased AP (Average Precision) and FPS by 10% and 12%, respectively, compared to v3. In v4, various deep learning techniques (WRC, CSP etc.) are used to improve performance, and the CSPNet-based backbone (CSPDarkNet53) is used. YOLOv5 [32] uses the same CSPNet-based backbone as YOLOv4. It is a PyTorch implementation, not Darknet, which is different from previous versions. Compared to v4, YOLOv5 is characterized by being able to configure and implement the environment more easily. YOLO v6 [33] is slightly harder to use in practice than YOLOv5, and there are not as many established paths and articles on how to actually use networks for training, deployment, and debugging. Starting with YOLOv6, it is possible to detect objects of various sizes, with the existing three scales increased to four. YOLOv7 [34] proposes a trainable bag-of-freebies method for real-time object detection that can significantly improve detection accuracy without increasing inference cost. In addition, it uses 'extend' and 'compound scaling' methods for real-time object detectors that can effectively utilize parameters and computation. We tested the models on an LS1028a board as an example of a lightweight embedded board. Figure 26 shows the time and fps measurement result with different versions of YOLO. YOLOv4 took 0.23 s and had 4.31 FPS. YOLOv5 took 0.228 s and had 4.38 FPS. YOLOv6 took 0.229 s and had 4.35 FPS. YOLOv7 took 0.23 s and had 4.2 FPS. The higher versions had better average time and fps. In addition, it can be seen that all versions above YOLOv3 showed improved results compared to the previous experiment.

Algorithms	YOLOv4	YOLOv5	YOLOv6	YOLOv7
Average time	0.23s	0.228s	0.229s	0.23s
Average fps	4.31	4.38	4.35	4.2

Figure 26. Time and FPS measurement of different versions of YOLO.

5. Conclusions and Discussion

This paper introduces an ROI masking method using semantic segmentation with ENet and an algorithm that can execute object recognition in real time on a lightweight embedded board using YOLO. It uses a webcam to receive real-time image input. Using an ENet model that has been trained to recognize only humans, image frames are converted to an appropriate size and then segmented. After binarizing the segmentation result and masking it for the ROI, the resulting images are used for object recognition with YOLO.

Our results show that using ENet to set the ROI improves accuracy significantly. The number of errors drops from 7 to 2.9. This algorithm can be judged valid because this increase in accuracy is achieved while increasing memory usage by only about two times, while power consumption is reduced from 32.8 W to 22.54 W when using ENet as the ROI setting. By testing the algorithm in several deep learning framework such as SSD, MobileNet, and Faster RCNN, we found that the average time required decreased by about 1.5 times, and the number of errors diminished to half. As a result of testing the different versions of YOLO developed thus far, the results for version 5 were the fastest at 4.38 FPS. Filtering the input image once using segmentation and then using the result to recognize an object increases the accuracy and reduces the required amount of computation. In addition, by dividing the code into two operations, the amount of computation can be further reduced. If the divided code is shared on different embedded boards to process images while communicating with each other, a lighter real-time image processing algorithm can be implemented.

Through this study, it has been found that the efficiency of object recognition can be greatly increased by using two deep learning models. It is expected that this method can be used for autonomous driving and IoT, which are fields that currently need to recognize people using object recognition. In addition, because it can be implemented on a much lighter and cheaper boards than the boards currently used in object recognition, it can be seen that the potential for grafting is high. Experimenting with different deep learning object recognition frameworks and different versions shows that this algorithm can be implemented flexibly. Therefore, we predict that the method devised here can be used with several deep learning and machine learning-based object recognition structures currently being studied.

In the future, research can be conducted into real-time object recognition based on deep learning to improve the accuracy in various environments and to optimize for operation on even lighter embedded boards. In addition, research on reducing the overflow that occurs during real-time image analysis by utilizing communication technologies such as socket communication can be studied. In addition, because the current object recognition deep learning algorithm was developed very rapidly in several ways, we plan to study it further in order to execute it more flexibly in various frameworks and languages according to the flow.

Author Contributions: H.Y. designed, implemented, and analyzed the entire model architecture; D.P. was the corresponding author. All authors have read and agreed to the published version of the manuscript.

Funding: This study was supported by the BK21 FOUR project funded by the Ministry of Education of Korea (4199990113966) and the Basic Science Research Program through the National Research Foundation of Korea (NRF) funded by the Ministry of Education (NRF-2018R1A6A1A03025109, 10%, NRF-2022R1I1A3069260, 20%) and the Ministry of Science and ICT (2020M3H2A1078119). This work was partly supported by an Institute of Information and Communications Technology Planning and Evaluation (IITP) grant funded by the Korean Government (MSIT) (No. 2021-0-00944, Metamorphic approach of unstructured validation/verification for analyzing binary code, 40%) and (No. 2022-0-00816, OpenAPI-based hw/sw platform for edge devices and cloud server, integrated with the on-demand code streaming engine powered by AI, 20%) and (No. 2022-0-01170, PIM Semiconductor Design Research Center, 10%). The EDA tool was supported by the IC Design Education Center (IDEC), Korea.

Data Availability Statement: Not applicable.

Conflicts of Interest: The authors declare no conflict of interest. The founding sponsors had no role in the design of the study, in the collection, analyses, or interpretation of data, in the writing of the manuscript, or in the decision to publish the results.

Abbreviations

The following abbreviations are used in this manuscript:

YOLO	You Look Only Once
ROI	Region Of Interest
ENet	Efficient Neural network
R-CNN	Regions with Convolutional Neuron Networks
FCN	Fully Convolutional Networks
CNN	Convolutional Neural Network
OpenCV	Open-Source Computer Vision
FPS	Frames Per Second

References

1. Lee, S.; Lee, D.; Choi, P.; Park, D. Efficient Power Reduction Technique of LiDAR Sensor for Controlling Detection Accuracy Based on Vehicle Speed. *IEMEK J. Embed. Syst. Appl.* **2020**, *15*, 215–225.
2. Lee, S.; Park, K.H.; Park, D. Communication-power overhead reduction method using template-based linear approximation in lightweight ecg measurement embedded device. *IEMEK J. Embed. Syst. Appl.* **2020**, *15*, 205–214.
3. Kim, J.; Kim, S. Autonomous-flight Drone Algorithm use Computer vision and GPS. *IEMEK J. Embed. Syst. Appl.* **2016**, *11*, 193–200.
4. Yogamani, S.; Hughes, C.; Horgan, J.; Sistu, G.; Varley, P.; O'Dea, D.; Uricár, M.; Milz, S.; Simon, M.; Amende, K.; et al. Woodscape: A multi-task, multi-camera fisheye dataset for autonomous driving. In Proceedings of the IEEE/CVF International Conference on Computer Vision, Seoul, Republic of Korea, 27 October 2019–2 November 2019; pp. 9308–9318.
5. Zhao, Z.Q.; Zheng, P.; Xu, S.T.; Wu, X. Object detection with deep learning: A review. *IEEE Trans. Neural Netw. Learn. Syst.* **2019**, *30*, 3212–3232. [CrossRef]
6. Huang, Y.; Li, Y.; Hu, X.; Ci, W. Lane detection based on inverse perspective transformation and Kalman filter. *KSII Trans. Internet Inf. Syst. (TIIS)* **2018**, *12*, 643–661.
7. Akhtar, M.N.; Saleh, J.M.; Awais, H.; Bakar, E.A. Map-Reduce based tipping point scheduler for parallel image processing. *Expert Syst. Appl.* **2020**, *139*, 112848. [CrossRef]
8. Wang, K.; Lin, L.; Yan, X.; Chen, Z.; Zhang, D.; Zhang, L. Cost-Effective Object Detection: Active Sample Mining With Switchable Selection Criteria. *IEEE Trans. Neural Netw. Learn. Syst.* **2019**, *30*, 834–850. [CrossRef]
9. He, K.; Gkioxari, G.; Dollár, P.; Girshick, R. Mask R-CNN. In Proceedings of the IEEE International Conference on Computer Vision, Venice, Italy, 22–29 October 2017; pp. 2961–2969.
10. Redmon, J.; Divvala, S.; Girshick, R.; Farhadi, A. You only look once: Unified, real-time object detection. In Proceedings of the IEEE Conference on Computer Vision and Pattern Recognition, Las Vegas, NV, USA, 27–30 June 2016; pp. 779–788.
11. Wang, P.; Chen, P.; Yuan, Y.; Liu, D.; Huang, Z.; Hou, X.; Cottrell, G. Understanding convolution for semantic segmentation. In Proceedings of the 2018 IEEE Winter Conference on Applications of Computer Vision (WACV), Lake Tahoe, NV, USA, 12–15 March 2018; pp. 1451–1460.
12. Dai, J.; Li, Y.; He, K.; Sun, J. R-fcn: Object detection via region-based fully convolutional networks. In Proceedings of the Advances in Neural Information Processing Systems 29 (NIPS 2016), Barcelona, Spain, 5–11 December 2016.
13. Paszke, A.; Chaurasia, A.; Kim, S.; Culurciello, E. Enet: A deep neural network architecture for real-time semantic segmentation. *arXiv* **2016**, arXiv:1606.02147.
14. He, K.; Zhang, X.; Ren, S.; Sun, J. Deep residual learning for image recognition. In Proceedings of the IEEE Conference on Computer Vision and Pattern Recognition, Las Vegas, NV, USA, 27–30 June 2016; pp. 770–778.
15. Yun, H.; Park, D. Mitigating Overflow of Object Detection Tasks Based on Masking Semantic Difference Region of Vision Snapshot for High Efficiency. In Proceedings of the 2022 International Conference on Artificial Intelligence in Information and Communication (ICAIIC), Jeju Island, Republic of Korea, 21–24 February 2022; pp. 138–140.
16. Kim, S.; Ji, Y.; Lee, K.B. An effective sign language learning with object detection based ROI segmentation. In Proceedings of the 2018 Second IEEE International Conference on Robotic Computing (IRC), Laguna Hills, CA, USA, 31 January 2018–2 February 2018; pp. 330–333.
17. Andriluka, M.; Pishchulin, L.; Gehler, P.; Schiele, B. 2d human pose estimation: New benchmark and state of the art analysis. In Proceedings of the IEEE Conference on Computer Vision and Pattern Recognition, Columbus, OH, USA, 23–28 June 2014; pp. 3686–3693.
18. Zhao, H.; Qi, X.; Shen, X.; Shi, J.; Jia, J. Icnet for real-time semantic segmentation on high-resolution images. In Proceedings of the European Conference on Computer Vision (ECCV), Munich, Germany, 8–14 September 2018; pp. 405–420.

19. Abadi, M.; Barham, P.; Chen, J.; Chen, Z.; Davis, A.; Dean, J.; Devin, M.; Ghemawat, S.; Irving, G.; Isard, M.; et al. TensorFlow: A System for Large-Scale Machine Learning. In Proceedings of the 12th USENIX Symposium on Operating Systems Design and Implementation (OSDI 16), Savannah, GA, USA, 2–4 November 2016; pp. 265–283.
20. Kim, D.G.; Park, Y.S.; Park, L.J.; Chung, T.Y. Developing of new a tensorflow tutorial model on machine learning: focusing on the Kaggle titanic dataset. *IEMEK J. Embed. Syst. Appl.* **2019**, *14*, 207–218.
21. Liu, W.; Anguelov, D.; Erhan, D.; Szegedy, C.; Reed, S.; Fu, C.Y.; Berg, A.C. Ssd: Single shot multibox detector. In Proceedings of the European Conference on Computer Vision, Amsterdam, The Netherlands, 11–14 October 2016; pp. 21–37.
22. Howard, A.G.; Zhu, M.; Chen, B.; Kalenichenko, D.; Wang, W.; Weyand, T.; Andreetto, M.; Adam, H. Mobilenets: Efficient convolutional neural networks for mobile vision applications. *arXiv* **2017**, arXiv:1704.04861.
23. Ren, S.; He, K.; Girshick, R.; Sun, J. Faster R-CNN: Towards real-time object detection with region proposal networks. In Proceedings of the Advances in Neural Information Processing Systems 28 (NIPS 2015), Montreal, Canada, 7–12 December 2015.
24. Trieu, T.H. GitHub Repository. Darkflow. Available online: https://github.com/thtrieu/darkflow (accessed on 10 June 2022).
25. Yun, H.; Park, D. Yolo-based Realtime Object Detection using Interleaved Redirection of Time-Multiplexed Streamline of Vision Snapshot for Lightweighted Embedded Processors. In Proceedings of the 2021 International Symposium on Intelligent Signal Processing and Communication Systems (ISPACS), Hualien City, Taiwan, 16–19 November 2021; pp. 1–2.
26. NXP. Layerscape LS1028A Family of Industrial Applications Processors. Available online: https://www.nxp.com/docs/en/fact-sheet/LS1028AFS.pdf (accessed on 15 July 2021).
27. NXP. Layerscape LX2160A Communications Processor. Available online: https://www.nxp.com/docs/en/fact-sheet/LX2160AFS.pdf (accessed on 15 July 2021).
28. Townsend, J.T. Theoretical analysis of an alphabetic confusion matrix. *Percept. Psychophys.* **1971**, *9*, 40–50. [CrossRef]
29. Redmon, J.; Farhadi, A. YOLO9000: Better, faster, stronger. In Proceedings of the IEEE Conference on Computer Vision and Pattern Recognition, Honolulu, HI, USA, 21–26 July 2017; pp. 7263–7271.
30. Ma, J.; Chen, L.; Gao, Z. Hardware implementation and optimization of tiny-YOLO network. In Proceedings of the International Forum on Digital TV and Wireless Multimedia Communications, Shanghai, China, 8–9 November 2017; pp. 224–234.
31. Bochkovskiy, A.; Wang, C.Y.; Liao, H.Y.M. Yolov4: Optimal speed and accuracy of object detection. *arXiv* **2020**, arXiv:2004.10934.
32. Jocher, G. ultralytics/yolov5: V6.0—YOLOv5n 'Nano' models, Roboflow integration; TensorFlow Export; OpenCV DNN Support. Available online: https://zenodo.org/record/5563715#.Y3LvP3ZBxdh (accessed on 5 May 2022)
33. Li, C.; Li, L.; Jiang, H.; Weng, K.; Geng, Y.; Li, L.; Ke, Z.; Li, Q.; Cheng, M.; Nie, W.; et al. YOLOv6: A single-stage object detection framework for industrial applications. *arXiv* **2022**, arXiv:2209.02976.
34. Wang, C.Y.; Bochkovskiy, A.; Liao, H.Y.M. YOLOv7: Trainable bag-of-freebies sets new state-of-the-art for real-time object detectors. *arXiv* **2022**, arXiv:2207.02696.

Article

Pantograph Detection Algorithm with Complex Background and External Disturbances

Ping Tan [1], Zhisheng Cui [1], Wenjian Lv [1], Xufeng Li [2], Jin Ding [1,*], Chuyuan Huang [3], Jien Ma [2] and Youtong Fang [2]

[1] School of Automation and Electrical Engineering, Zhejiang University of Science and Technology, Hangzhou 310023, China
[2] College of Electrical Engineering, Zhejiang University, Hangzhou 310027, China
[3] Chinese-German Institute for Applied Engineering, Zhejiang University of Science and Technology, Hangzhou 310023, China
* Correspondence: jding@zust.edu.cn

Abstract: As an important equipment for high-speed railway (HSR) to obtain electric power from outside, the state of the pantograph will directly affect the operation safety of HSR. In order to solve the problems that the current pantograph detection method is easily affected by the environment, cannot effectively deal with the interference of external scenes, has a low accuracy rate and can hardly meet the actual operation requirements of HSR, this study proposes a pantograph detection algorithm. The algorithm mainly includes three parts: the first is to use you only look once (YOLO) V4 to detect and locate the pantograph region in real-time; the second is the blur and dirt detection algorithm for the external interference directly affecting the high-speed camera (HSC), which leads to the pantograph not being detected; the last is the complex background detection algorithm for the external complex scene "overlapping" with the pantograph when imaging, which leads to the pantograph not being recognized effectively. The dirt and blur detection algorithm combined with blob detection and improved Brenner method can accurately evaluate the dirt or blur of HSC, and the complex background detection algorithm based on grayscale and vertical projection can greatly reduce the external scene interference during HSR operation. The algorithm proposed in this study was analyzed and studied on a large number of video samples of HSR operation, and the precision on three different test samples reached 99.92%, 99.90% and 99.98%, respectively. Experimental results show that the algorithm proposed in this study has strong environmental adaptability and can effectively overcome the effects of complex background and external interference on pantograph detection, and has high practical application value.

Keywords: high-speed railway; object detection; blob detection; EOR-Brenner; blur and dirt; complex background

Citation: Tan, P.; Cui, Z.; Lv, W.; Li, X.; Ding, J.; Huang, C.; Ma, J.; Fang, Y. Pantograph Detection Algorithm with Complex Background and External Disturbances. *Sensors* **2022**, *22*, 8425. https://doi.org/10.3390/s22218425

Academic Editors: Sylvain Girard and Enrico Meli

Received: 21 August 2022
Accepted: 28 October 2022
Published: 2 November 2022

Publisher's Note: MDPI stays neutral with regard to jurisdictional claims in published maps and institutional affiliations.

1. Introduction

As an important part of the pantograph-catenary system (PCS), the pantograph is a special current-receiving device installed on the roof of the high-speed railway (HSR). When the pantograph is raised, it transmits power from the traction substation to the HSR through the friction between the pantograph and the contact network, thus providing the power required for the operation of the HSR. Once a pantograph failure occurs, it will directly affect the operational safety of HSR [1–3]. Therefore, the current pantograph status must be accurately assessed through real-time detection of pantographs to ensure the safety and stability of HSR operation. The PCS is shown in Figure 1.

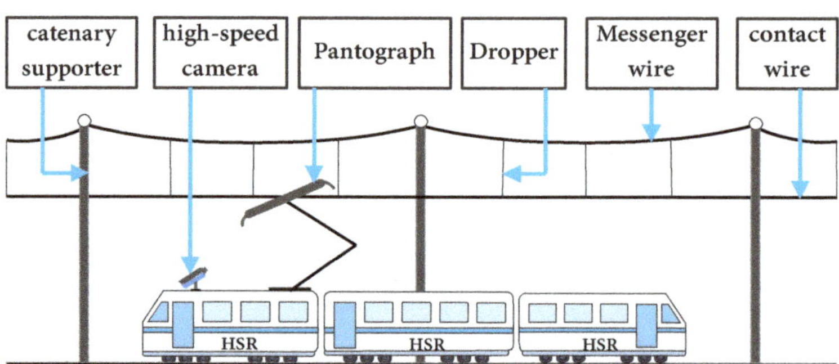

Figure 1. Schematic of PCS.

There are two main models of HSR in actual operation, the speed of the two models of HSR is usually 150–300 km/h when they are running stably, but the images captured by the high-speed cameras (HSC) equipped with the two models of HSR are slightly different. One is the image captured by HSR-A as shown in the left image in Figure 2, and the other is the image captured by HSR-B as shown in the right image in Figure 2. It is worth mentioning that there are some Chinese messages in the images captured by the HSC in Figure 2, which contain the basic information of the vehicle and the time information and do not affect the reader's understanding of this paper. The same is true for the images captured by the relevant HSC that appear subsequently in the paper.

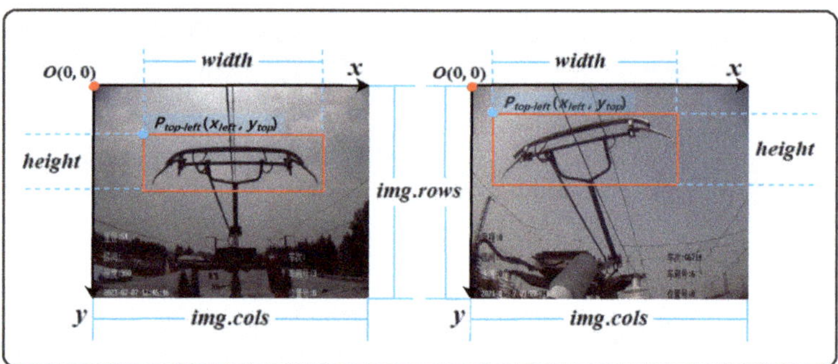

Figure 2. HSC footage of pantographs.

Although the two models of HSR are equipped with different angles of HSC, they both have a frame rate of 25 FPS. Therefore, regardless of the operating speed of HSR, the HSC can only capture 25 pantograph images per second, so the algorithm must process at least 25 images captured by the HSC per second to meet the real-time requirement. The region corresponding to the red rectangle in Figure 2 is the region of interest (ROI), and the pantograph in the ROI is the main research object of this study.

In the current pantograph detection method, Refs. [4,5] proposed the use of Catenary and Pantograph Video Monitor (CPVM-5C) System for pantograph detection, but in the 5C system the camera is generally installed at the HSR exit, which cannot detect and monitor the running HSR in real time. Refs. [6–8] proposed to extract the edges of pantographs by improved edge detection, wavelet transform, hough transform, etc., so as to realize the evaluation of pantographs, but this is essentially based on the traditional image processing method, which is only applicable to pantograph detection when the overall image is clear and the background is single, which is limited and difficult to meet the complex situation when the HSR is actually running. Refs. [9–11] proposed to achieve real-time pantograph detection by simply using a certain improved neural network, whose detection results are entirely given by the neural network. This method relies heavily on a large number of data sets for support, and is prone to a large number of false alarms when the training set is not rich enough in samples. The data set of certain complex scenes in the operation of HSR is difficult to obtain, so it is difficult to build a model that covers a large number of rich scene samples under training, which makes a large interference to the detection results when disturbed. Refs. [12–15] and others combine deep learning and image processing to greatly improve the stability of pantograph detection by a single reliance on neural networks, but there are still major limitations in complex scenes. The proposed methods of [16–18] are not very practical for complex scenes and external interference, and the complex scenes that can be overcome are very limited.

In the actual operation of HSR, it is often faced with various complex environments and changing scenarios. Even for HSR running on the same line, there may be huge differences in the scenarios encountered in different time periods. This difference is caused by multiple factors, which is irregular and difficult to predict. Because the occurrence of these scenes is full of randomness, resulting in a sample set for training neural networks that cannot cover all situations in all complex scenes and environments. With limited samples, methods to improve detection accuracy by improving certain neural networks do not fundamentally address the large number of pantograph state false positives in such scenarios, and cannot really address the impact of complex scenarios in the actual operation of HSR. Therefore, this paper focuses on filtering and detecting these complex scenes and external interference by designing algorithms, so as to achieve a method more in line with the actual operation of HSR and more widely applicable, reducing or even eliminating these scenes for neural network real-time detection of a pantograph's impact.

2. YOLO V4 Locates the Pantograph Region

The Alexey-proposed You Only Look Once (YOLO) V4 is a huge upgrade to the one-stage detector in the field of object detection [19]. Compared with the previous version of YOLO, YOLO V4 replaces the backbone network from the original darknet53 to CPSdarknet53 on the basis of YOLO V3, which makes YOLO V4 effectively reduce the amount of computation and improve the learning ability. Meanwhile, YOLO V4 replaces spatial pyramid pooling (SPP) with feature pyramid networks (FPN), which splices feature maps at different scales and increases the receptive field of the model, enabling YOLO V4 to extract more details.

Average Precision (AP) and Mean Average Precision (mAP) are important metrics to measure the performance of the target detection algorithm, while AP-50 and AP-75 are the AP values when the corresponding Intersection over Union (IoU) thresholds are set to 0.5 and 0.75. The performance of YOLO V4 and current mainstream object detection algorithms on two datasets, Visual Object Classes (VOC) and Common Objects in Context (COCO), is shown in Figure 3.

Figure 3 shows that the YOLO V4 has clear advantages in all aspects. Alexey had pointed out that the YOLO V4 was the most advanced detector at that time, and even now it still seems that the YOLO V4 has great advantages and performance [19]. Therefore, YOLO V4 is used to locate the pantograph region in this study, and the located pantograph region is passed into the subsequent algorithm. The overall algorithm flow for locating the pantograph region using YOLO V4 is shown in Figure 4.

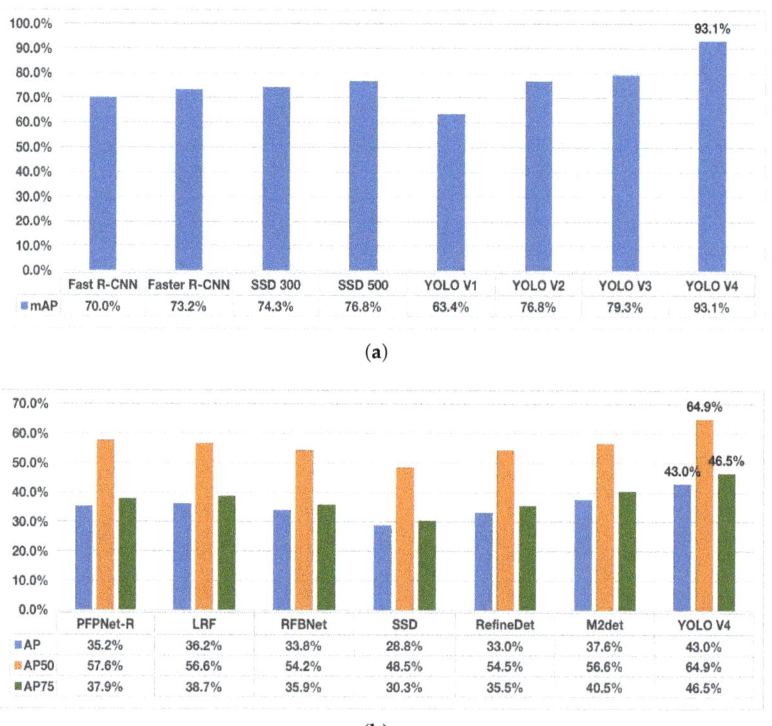

(a)

(b)

Figure 3. Comparison of YOLO V4 with other mainstream neural networks [20–32]. (**a**) Test results on VOC2007 + VOC2012. (**b**) Test results on the COCO dataset.

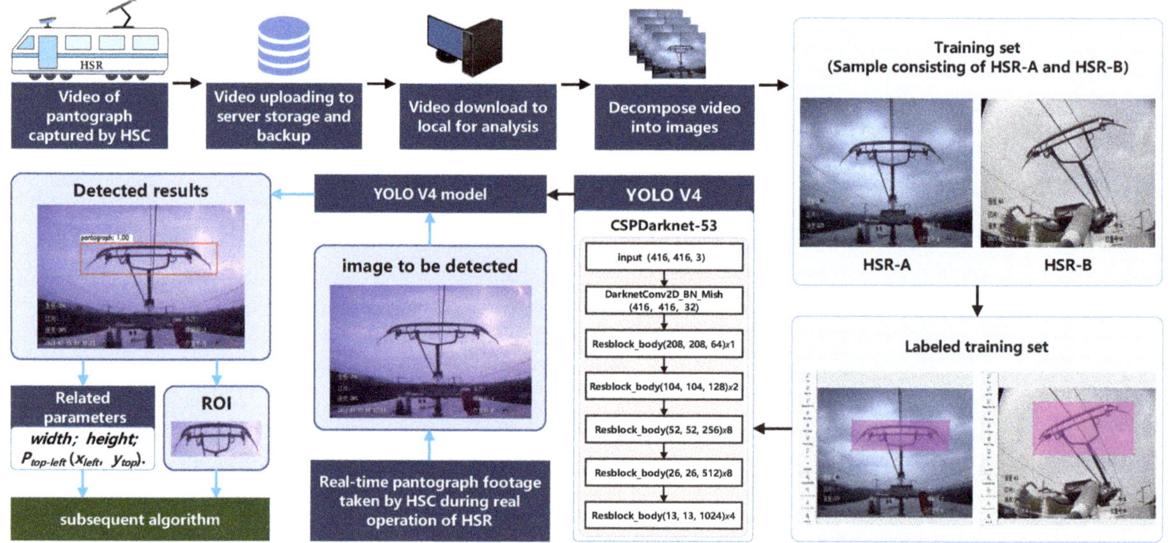

Figure 4. YOLO V4 overall algorithm process.

3. HSC Blur and Dirt Detection Algorithm

3.1. Blurry HSC Screen and Dirty HSC Screen

During the operation of HSR, the HSC is always exposed to the outside of the car, which makes the HSC extremely vulnerable to interference from the outside. The external interference affecting the HSC is mainly divided into two kinds: one is the influence of rain on the pantograph during rainy days, and the other is the influence of the dirt attached to the HSC lens on the pantograph.

3.1.1. Rainwater

HSR operation needs to face very complicated weather conditions, especially in rainy days, rainwater will directly affect the imaging of HSC. Figure 5 illustrates the different degrees of impact of rain on the HSR-A and HSR-B. when HSR is running at high speed, rainwater tends to cause blurring of the HSC imaging, making the captured pantographs unclear and thus causing the YOLO V4 to incorrectly assess the pantographs.

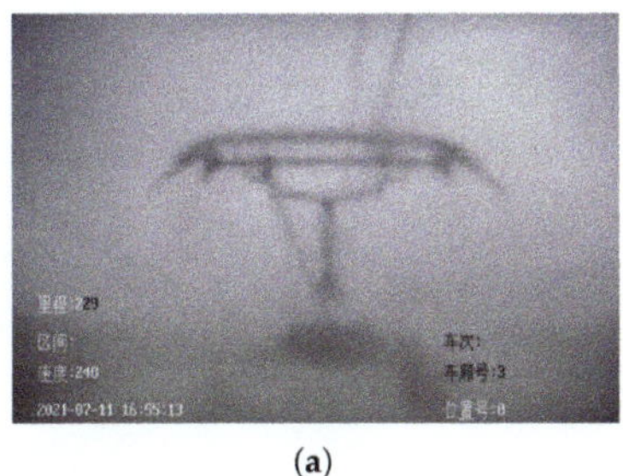

(**a**)

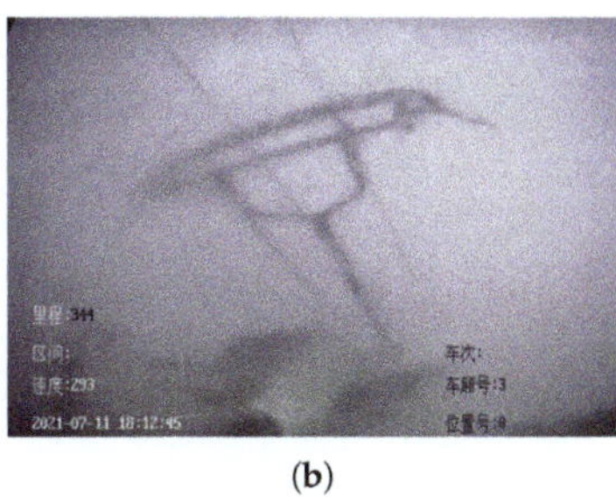

(**b**)

Figure 5. Blurred HSC imaging caused by rainwater. (**a**) HSR-A. (**b**) HSR-B.

3.1.2. Dirty

The lens dirt attached to the HSC can generally only be removed by manual cleaning. As shown in Figure 6, during the period from the time when the lens is dirty to before the dirt is artificially cleaned, the dirty lens will continue to affect the overall evaluation of the pantograph by YOLO V4.

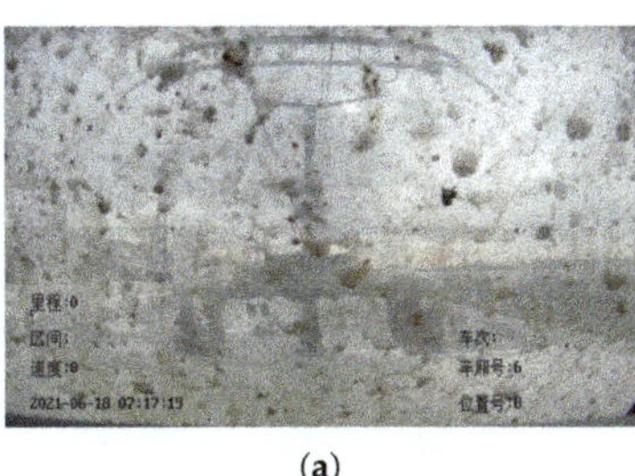

(**a**)

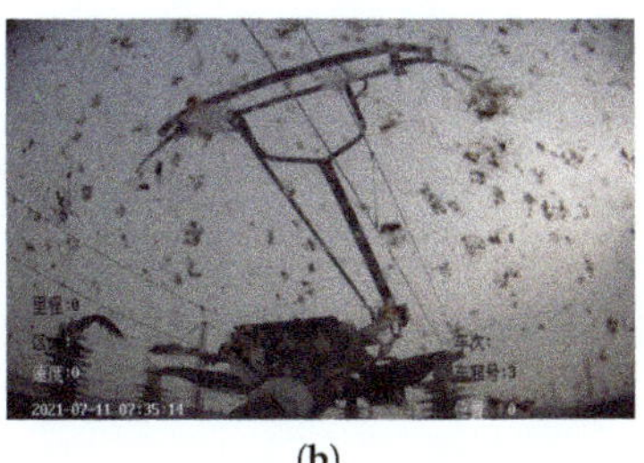

(**b**)

Figure 6. The HSC lens has a lot of dirt attached to it. (**a**) HSR-A. (**b**) HSR-B.

3.2. External Factors Cause YOLO V4 to Fail to Locate the Pantograph

When YOLO V4 cannot locate the pantograph due to external interference, the approximate position of the pantograph in the current screen can be inferred from the pantograph position that was determined in the previous normal screen. When YOLO V4 locates the pantograph area, it only needs to obtain four parameters of the bounding box in Figure 2 to achieve its accurate positioning. These four parameters are the horizontal coordinates (x_{left}) and vertical coordinates (y_{top}) of the point ($P_{top-left}$) in the upper left corner of the bounding box, and the width and height of the pantograph. The variation of the four parameters of the bounding box positioned by YOLO during normal operation of HSR of two different models is shown in Figure 7.

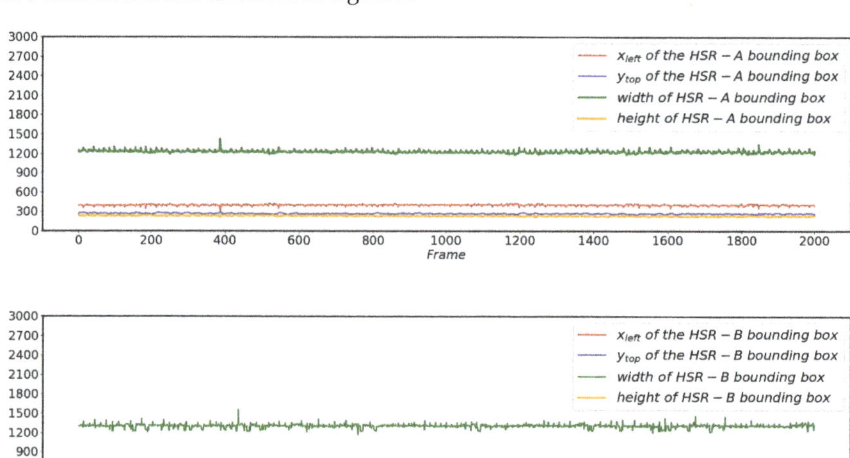

Figure 7. Changes of the four parameters of the bounding box when YOLO V4 is positioned normally without external interference.

As can be seen from Figure 7, whether it is HSR-A or HSR-B, when its normal operation is not disturbed by external scenes, the pantograph region positioned by YOLO V4 is always relatively fixed, although there is a small range of jitter. This small-scale jitter is caused by a combination of factors such as the bumps during the operation of the HSR and the force changes between the pantograph and the catenary. This jitter does not affect the approximate position of the pantograph in the image, so when the YOLO V4 is unable to locate the pantograph area due to external interference, the approximate position of the pantograph in the current frame can be inferred from the coordinate information obtained from the previous frame, and subsequent analysis can be performed.

3.3. Improved Image Sharpness Evaluation Algorithm

Brenner algorithm is a classical blur detection algorithm [33], which finally achieves the evaluation of image sharpness by accumulating the square of the grayscale difference between two pixel points. Since the gray value of the image at the focal position changes significantly compared with the telefocused image, and the image at the focal position has more edge information, a more accurate judgment of the sharpness of the image can be made using this method. However, the traditional Brenner algorithm cannot meet the complex scene changes and variable external disturbances that need to be faced during the operation of high speed rail, so this paper proposes the emphasize object region-Brenner (EOR-Brenner) algorithm combined with the pantograph region localized by YOLO V4. The principle of EOR-Brenner is shown in Equation (1).

$$
\begin{aligned}
F &= k_1 F_{IMG} + k_2 F_{ROI} \\
&= k_1 \sum_{x=0}^{img.cols-3} \sum_{y=0}^{img.rows-1} [f(x+2,y) - f(x,y)]^2 \\
&\quad + k_2 \sum_{x=x_{left}}^{x_{left}+width} \sum_{y=y_{top}}^{y_{top}+height} [f(x+2,y) - f(x,y)]^2
\end{aligned}
\tag{1}
$$

where x is the horizontal coordinate of a pixel point, y is the vertical coordinate of a pixel point, $f(x, y)$ is the gray value of the pixel point, F_{IMG} and F_{ROI} are the sharpness results of the corresponding region. k_1 and k_2 are the weights of the corresponding region, and F is the final result of the improved Brenner algorithm.

Although the ROI occupies a relatively small area of the whole image, the pantograph, as the key research object, should be given a higher weight to the area where it is located. In this study, we recommend that k_1 can be 2 or 4 times of k_2, and the specific choice should be made flexibly according to the actual operation line of HSR. After the values of k_1 and k_2 are determined, the appropriate threshold (λ) is selected based on the calculated EOR-Brenner to achieve the differentiation and detection of clear and blurred images.

As shown in (2), when the final result F of EOR-Brenner is higher than the set threshold (λ), the image captured by the current HSC is considered to be clear. If the pantograph cannot be detected or is detected as abnormal at this time, it can be assumed that the current detection result is not affected by the blurring of the HSC screen. However, there are still two situations: (1) the current pantograph is in normal state, although it is not affected by the blurred screen, but it may be disturbed by other external environment such as complex background, which leads to the normal pantograph being undetectable or the pantograph is incorrectly detected as abnormal. (2) The pantograph is really abnormal. At this time, it is necessary to further evaluate the real state of the pantograph through the subsequent algorithm, and finally realize the accurate detection of the real state of the pantograph.

$$\begin{cases} Clear \quad image, & F > \lambda \\ Blurred \quad image, & F < \lambda \end{cases} \tag{2}$$

3.4. Blob Detection Algorithm Detects Screen Dirt

When dirt is attached to a HSC, it is very easy to form blobs. Blobs caused by dirt have different areas, convexity, circularity and inertia rates, so these attributes can be used to detect and filter the blobs [34–37], and the number of blobs can ultimately determine whether the HSC is dirty or not.

The area of the blob (S) reflects the size of the detected blob, while the circularity derived from the area of the blob (S) and the corresponding perimeter (C) reflects the degree to which the detected spot is close to a circle, and the calculation of the circularity is shown in Equation (3):

$$Value_{circularity} = \frac{4\pi S}{C^2} \tag{3}$$

The convexity reflects the degree of concavity of the blob. The convexity of the blob can be obtained from the area of the blob (S) and the area of the convex hull (H) of the blob, which is calculated as shown in Equation (4):

$$Value_{convexity} = \frac{S}{H} \tag{4}$$

The inertia rate also reflects the shape of the blob. If an image is represented by $f(x, y)$, then the moments of the image can be expressed by the Equation (5)

$$M_{ij} = \sum_{x} \sum_{y} x^i y^j f(x, y) \tag{5}$$

For a binary image, the zero-order moment M_{00} is equal to its area, so its center of mass is as shown in Equation (6):

$$\{\bar{x}, \bar{y}\} = \left\{ \frac{M_{10}}{M_{00}}, \frac{M_{01}}{M_{00}} \right\} \tag{6}$$

The central moment of the image is defined as shown in Equation (7):

$$\mu_{pq} = \sum_x \sum_y (x - \bar{x})^p (y - \bar{y})^q f(x,y) \tag{7}$$

If only second-order central moments are considered, the image is exactly equivalent to an ellipse with a defined size, orientation and eccentricity, centered at the image center of mass and with constant radiality. The covariance moments of the image are shown in Equation (8):

$$\text{cov}[f(x,y)] = \begin{bmatrix} \mu'_{20} & \mu'_{11} \\ \mu'_{11} & \mu'_{02} \end{bmatrix} = \begin{bmatrix} \frac{\mu_{20}}{\mu_{00}} & \frac{\mu_{11}}{\mu_{00}} \\ \frac{\mu_{11}}{\mu_{00}} & \frac{\mu_{02}}{\mu_{00}} \end{bmatrix} \tag{8}$$

The two eigenvalues λ_1 and λ_2 of this matrix correspond to the long and short axes of the image intensity (i.e., the ellipse). Then λ_1 and λ_2 can be expressed by the Equation (9):

$$\begin{aligned} \lambda_1 &= \frac{\mu'_{20} + \mu'_{02}}{2} + \frac{\sqrt{4\mu'^2_{11} + (\mu'_{20} - \mu'_{02})^2}}{2} \\ \lambda_2 &= \frac{\mu'_{20} + \mu'_{02}}{2} - \frac{\sqrt{4\mu'^2_{11} + (\mu'_{20} - \mu'_{02})^2}}{2} \end{aligned} \tag{9}$$

The final inertia rate is obtained as shown in Equation (10):

$$\begin{aligned} Value_{inertia} = \frac{\lambda_2}{\lambda_1} &= \frac{\mu'_{20} + \mu'_{02} - \sqrt{4\mu'^2_{11} + (\mu'_{20} - \mu'_{02})^2}}{\mu'_{20} + \mu'_{02} + \sqrt{4\mu'^2_{11} + (\mu'_{20} - \mu'_{02})^2}} \\ &= \frac{\mu_{20} + \mu_{02} - \sqrt{4\mu^2_{11} + (\mu_{20} - \mu_{02})^2}}{\mu_{20} + \mu_{02} + \sqrt{4\mu^2_{11} + (\mu_{20} - \mu_{02})^2}} \end{aligned} \tag{10}$$

The final selection of the number of blobs is achieved by the area, convexity, circularity and inertia rate of the blobs, and when the final number of detected blobs is greater than the set threshold, it can be inferred that the HSC surface is attached to the dirty at this time, so as to achieve the detection of HSC dirty. For the case shown in Figure 6 the final detection result is shown in Figure 8.

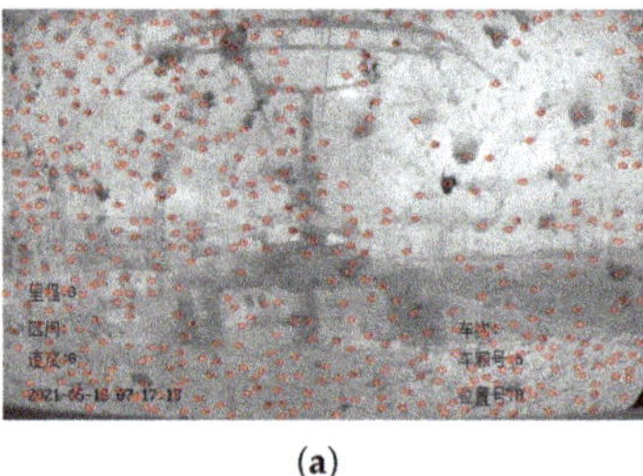
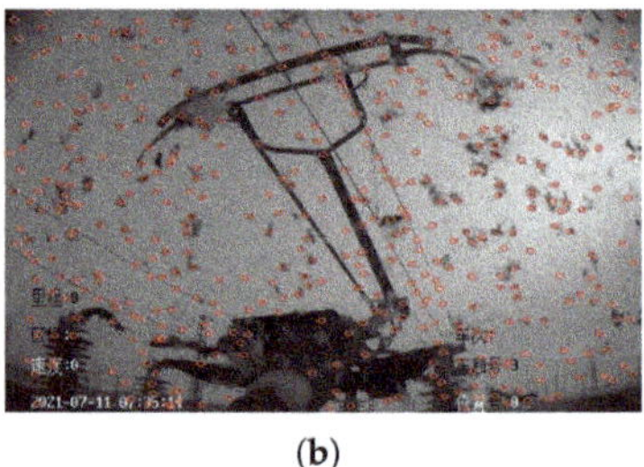

(a)   (b)

Figure 8. The HSC Screen dirty detection results. (**a**) HSR-A. (**b**) HSR-B.

3.5. Overall Process of HSC Blur and Dirt Detection Algorithm

As shown in Figure 9, the number of blobs in the current frame is first detected by the blob detection algorithm, and when the number is greater than the set threshold it is determined that the reason why YOLO V4 cannot achieve positioning in the current frame is due to dirt, and if the number of detected spots is less than the threshold value, the EOR-Brenner is used to evaluate whether the current frame is blurred or not. Finally correctly evaluate whether the pantograph detection abnormality in the current frame or the pantograph cannot be detected is caused by the dirty and blurred HSC.

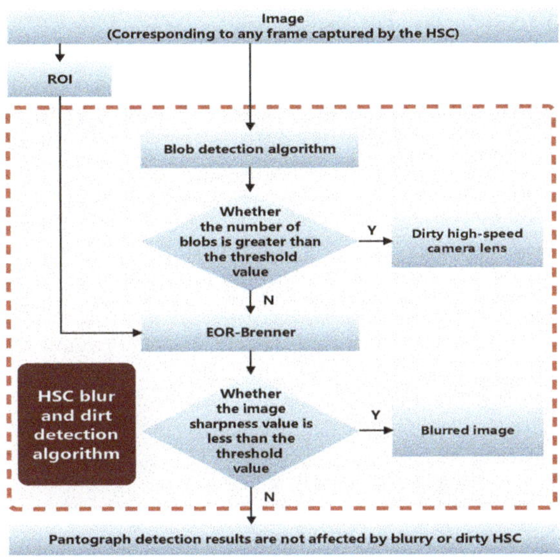

Figure 9. HSC blur and dirt detection algorithm process flow chart.

4. HSR Complex Background Detection Algorithm

4.1. The Complex Background That HSR Needs to Face

HSR often needs to face a large number of external scene changes and variable terrain, environment and other influences during actual operation. These external scenes and terrain, environment, etc. can directly affect the algorithm's correct assessment of the real state of the pantograph, and thus a large number of false alarms occur. Compared with blur and dirt, which directly affect the HSC and thus affect the detection of pantographs, when these external scenes and terrain environments affect the detection of pantographs, the images captured by the HSC are still very clear and free of blobs, but their impact on pantograph detection is mainly due to the HSC imaging when these external disturbances and pantograph "overlap" together, thus causing a large number of false alarms on the pantograph state. In this study, we refer to this type of interference as the "complex background", and the common complex backgrounds are catenary support devices, the sun, bridges, tunnels, and platforms of HSR.

In this study, we propose a HSR complex background detection algorithm to achieve accurate detection of these complex scenes during the operation of HSR, so as to exclude the influence of these complex background on the pantograph state evaluation.

4.1.1. Catenary Support Devices

As an extremely important part of the whole huge HSR system, the catenary support device not only plays the role of electrical insulation, but also bears a certain mechanical load. The contact network support device, as the most frequently appearing background, as shown in Figure 10 will often affect the normal detection of pantographs.

(a)   (b)

Figure 10. Catenary support device affects pantograph detection. (**a**) HSR-A. (**b**) HSR-B.

4.1.2. Sun

As shown in Figure 11, when the sun appears in the pantograph imaging region, the strong light causes a "partial absence"-like phenomenon in the pantograph.

 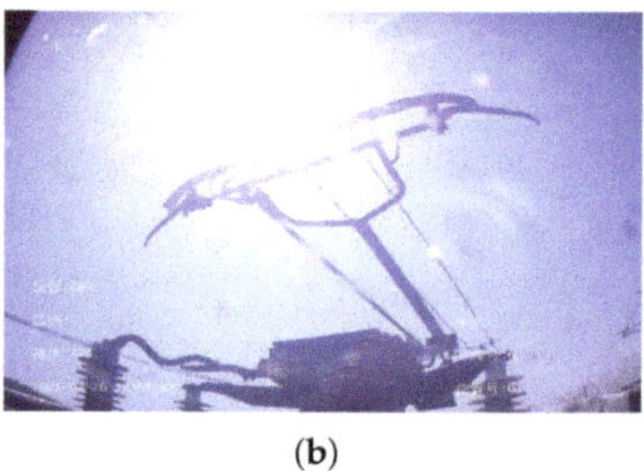

(a)   (b)

Figure 11. Sun affects pantograph detection. (**a**) HSR-A. (**b**) HSR-B.

4.1.3. Bridge

Due to the complex geographical environment, when two areas are separated by rivers, only special or mixed-use bridges can be built over the rivers to provide HSR access. In more and more cities, numerous viaducts are being built to provide access to HSR. When the HSR crosses the bridge, it directly affects the detection and positioning of the pantographs. The effect of bridges on pantographs is shown in the Figure 12.

(a)   (b)

Figure 12. Bride affects pantograph detection. (**a**) HSR-A. (**b**) HSR-B.

4.1.4. Tunnel

The presence of the tunnel greatly reduces the travel time and shortens the mileage between the two areas. Figure 13 shows the different images captured by the HSC before and after the HSR enters the tunnel. When the HSR enters the tunnel and runs stably, as shown in Figure 13c, the normal monitoring of the pantograph can still be achieved at this time because the fill light on the HSR is turned on. However, as shown in Figure 13b and Figure 13d, the dramatic light changes during the short period of time when the HSR enters and leaves the tunnel will cause the neural network to fail to achieve accurate positioning and detection of the pantographs when entering and leaving the tunnel.

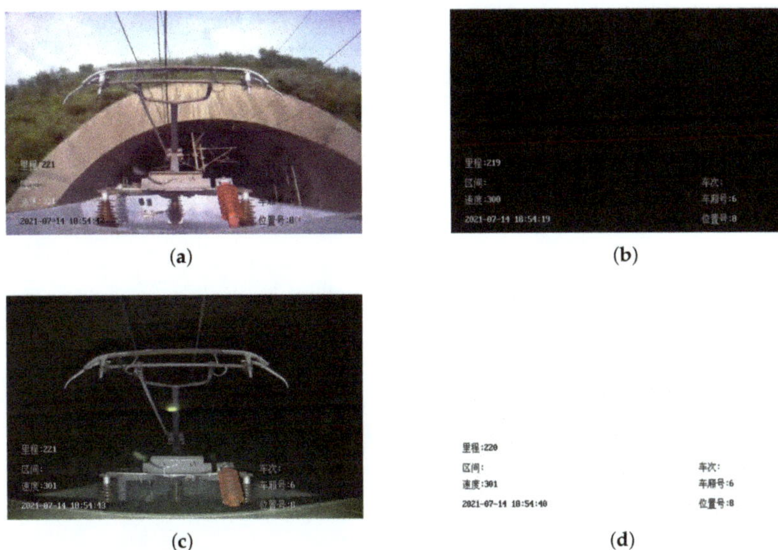

Figure 13. Tunnels affects pantograph detection. (**a**) Before the HSR enters the tunnel. (**b**) The moment the HSR enters the tunnel. (**c**) After the fill light is turned on, the HSR runs stably in the tunnel. (**d**) The moment the HSR exits the tunnel.

4.1.5. Platform

As shown in Figure 14, when the HSR drives into the platform, the platform will partially overlap with the pantograph region, which affects YOLO's positioning and detection of the pantograph, thus causing a large number of false alarms of the pantograph status by YOLO in the platform.

Figure 14. Platform affects pantograph detection. (**a**) HSR-A. (**b**) HSR-B.

4.2. Tunnel Detection Algorithm Based on the Overall Average Grayscale of the Image

For such false alarms caused by drastic changes in light over a short period of time that cause YOLO to be unable to detect and locate the pantograph for a short period of time, they can be excluded by the grayscale change rule of the image. The average grayscale calculation method of the image is shown in Equation (11):

$$\bar{g} = \frac{\sum_{i=0}^{img.cols-1} \sum_{j=0}^{img.rows-1} P(i,j)}{img.cols * img.rows} \tag{11}$$

where $P(x,y)$ is the grayscale of the corresponding pixel point, *img.rows* is the height of the image and *img.cols* is the width of the image.

When the pantograph is running in a relatively clear and clean background, the image corresponding to each frame will cause the average grayscale of the image to fluctuate in a small range with the continuous operation of the HSR and the continuous change of the scene, but there will not be a large change in the average grayscale. Figure 15 shows the change in the average grayscale of the images taken by the HSC before and after the different cars enter and exit the tunnel.

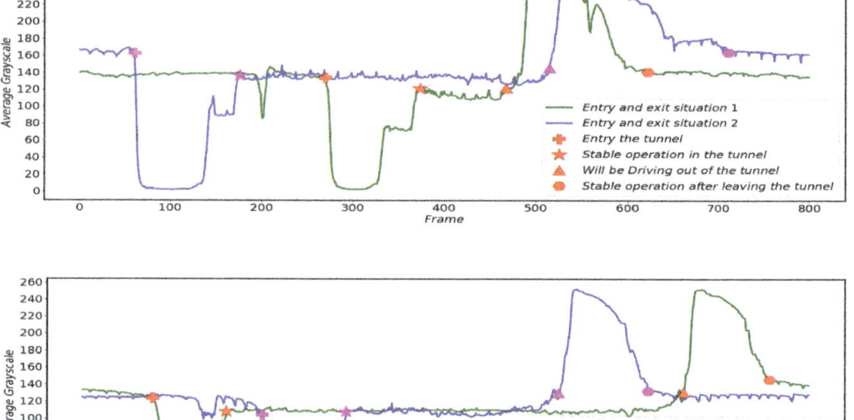

Figure 15. Average grayscale variation of images of HSR-A (**top**) and HSR-B (**bottom**) when driving into different tunnels.

As can be seen from Figure 15, when the HSR is running normally outside the tunnel, the average grayscale of the image only fluctuates in a very small range, and basically remains relatively stable. When the HSR enters the tunnel, the average gray value of the captured image drops to about 5 (as shown in Figure 13b, the image is basically black) because the fill light is not yet turned on and the light inside and outside the tunnel changes drastically. As the fill light is turned on, after a short period of time to adapt to the HSR will remain in a stable state in the tunnel and continue to travel, the average gray scale of the image will remain relatively stable again (as shown in Figure 13d, the image is basically all white) and the time of the HSR in the tunnel is determined by the speed of the HSR and the length of the tunnel. When the HSR out of the tunnel, due to run from a relatively dark environment to a bright environment, the HSC overexposure phenomenon will occur. At this time the average gray scale of the HSC captured by the image will jump to close to 250 or so.

4.3. Sun Detection Algorithm Based on Local Average Grayscale of Image Pantograph Region

The influence of the sun on the HSR is full of uncertainty. We cannot accurately predict that a HSR happens to pass by at a certain time on a certain line, and the sun also happens to appear in the pantograph imaging region of the HSR at this time, and affect YOLO's assessment of the pantograph state. Moreover, not all suns are as jealous of pantograph detection as shown in Figure 11. Figure 16 shows the situation where the sun appears in some images taken by HSC, but the sun does not affect YOLO's detection of the pantograph region.

(a) **(b)** **(c)** **(d)** **(e)** **(f)**

Figure 16. Sun did not affect YOLO detection of pantographs in HSR-A and HSR-B. (**a**) Case I. (**b**) Case II. (**c**) Case III. (**d**) Case IV. (**e**) Case V. (**f**) Case VI.

The screen of the corresponding scene in Figure 16 after the high speed rail leaves the area affected by the sun is shown in Figure 17. Furthermore, the average grayscale of the corresponding scenes in Figures 16 and 17 is shown in Figure 18.

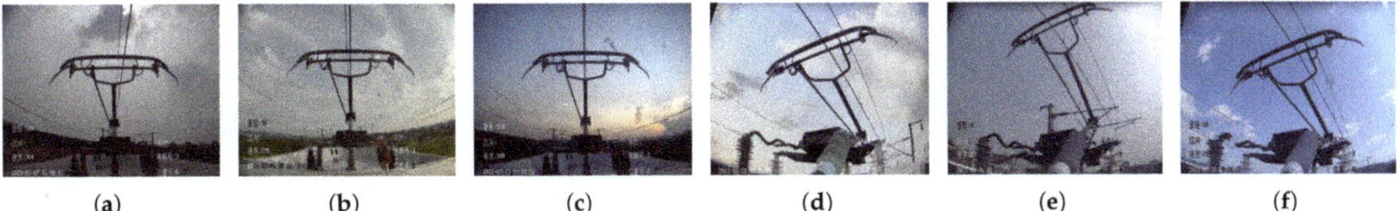

(a) **(b)** **(c)** **(d)** **(e)** **(f)**

Figure 17. The corresponding HSC in Figure 16 captures the scene without the sun in the frame. (**a**) Case I. (**b**) Case II. (**c**) Case III. (**d**) Case IV. (**e**) Case V. (**f**) Case VI.

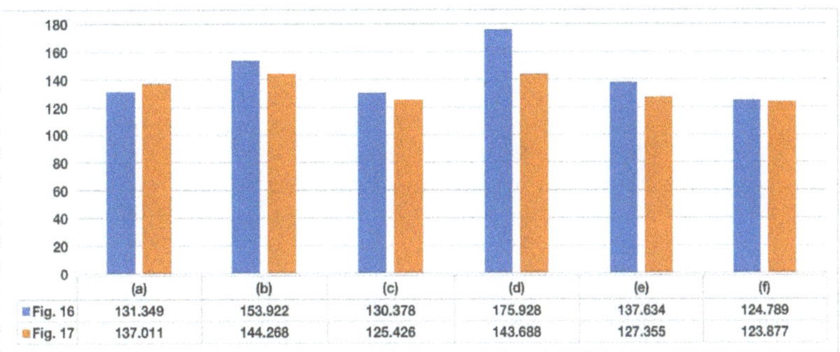

	(a)	(b)	(c)	(d)	(e)	(f)
Fig. 16	131.349	153.922	130.378	175.928	137.634	124.789
Fig. 17	137.011	144.268	125.426	143.688	127.355	123.877

Figure 18. Average grayscale comparison.

It can be found that the overall average grayscale of the image is not necessarily increased after the sun appears in the image captured by the HSC. However, when the sun affects the detection of pantographs, it will definitely cause an increase in the average grayscale of ROI. When the sun is not present the difference between the overall image and the average grayscale of the ROI is not significant, but once the sun affects the pantograph, it will definitely cause a large difference between the two. Using this unique difference, it is possible to determine whether the pantograph is detected as anomalous in the current image due to the sun. When the sun affects the pantograph detection, the average grayscale change of the overall image and ROI and the corresponding difference between the two average gray levels are shown in Figure 19.

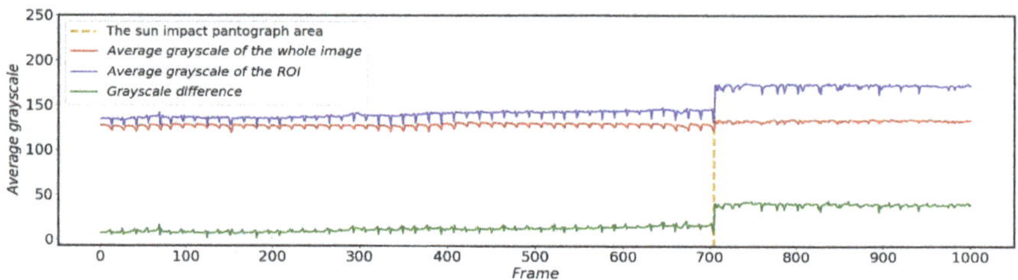

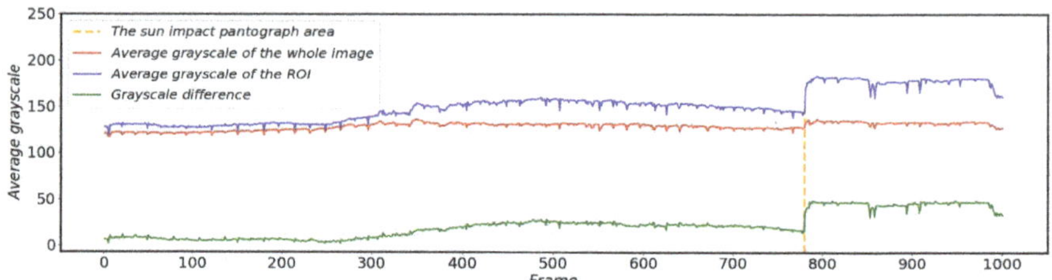

Figure 19. Average grayscale variation in the corresponding areas of HSR-A (**top**) and HSR-B (**bottom**) during sun influence pantograph detection.

4.4. Background Detection Algorithm for Catenary Support Devices, Bridges, and Platforms Based on Vertical Projection

Catenary support devices, bridges, and platforms do not have an excessive effect on the average grayscale of the images captured by the HSC, so for these three common external disturbances, the choice was made to eliminate the relevant interference by using vertical projection. As shown in Figure 20a, based on the ROI positioned by YOLO V4, the left region of interest (L-ROI) and right region of interest (R-ROI) can be positioned. Firstly, the image captured by the HSC is binarized to highlight the object to be studied, and the result of binarization is shown in Figure 20b. Then the binary image is passed through the image to reduce the interference in the image by the opening operation, and the image after the opening operation is shown in Figure 20c. Finally, the vertical projection of the L-ROI, ROI, and R-ROI regions is calculated by the result of the open operation as shown in Figure 21, where the height of the white region of the vertical projection reflects the number of pixels in the white region on the corresponding horizontal coordinates in the binary image.

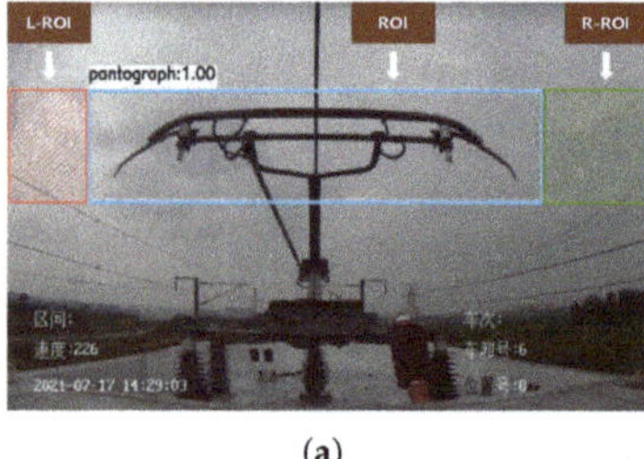

(**a**)

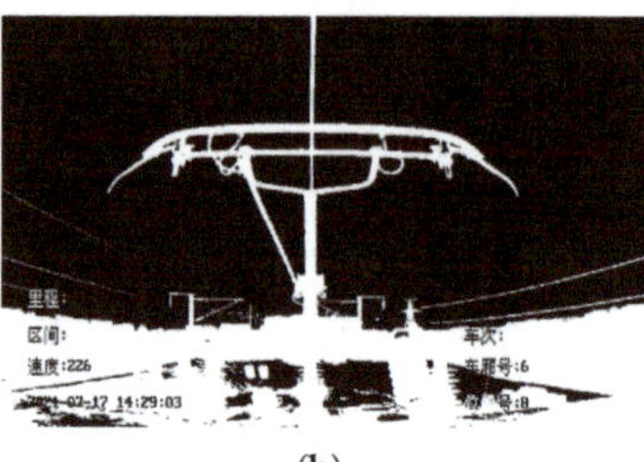
(**b**)

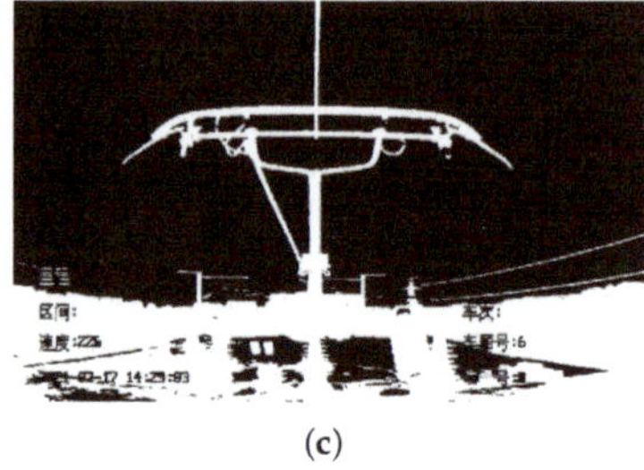
(**c**)

Figure 20. Image binarization and opening operations. (**a**) L-ROI, ROI and R-ROI. (**b**) Binary image. (**c**) Binary image after opening operation.

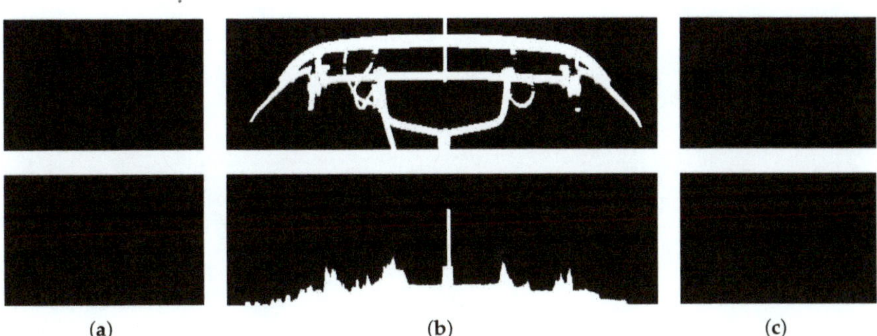

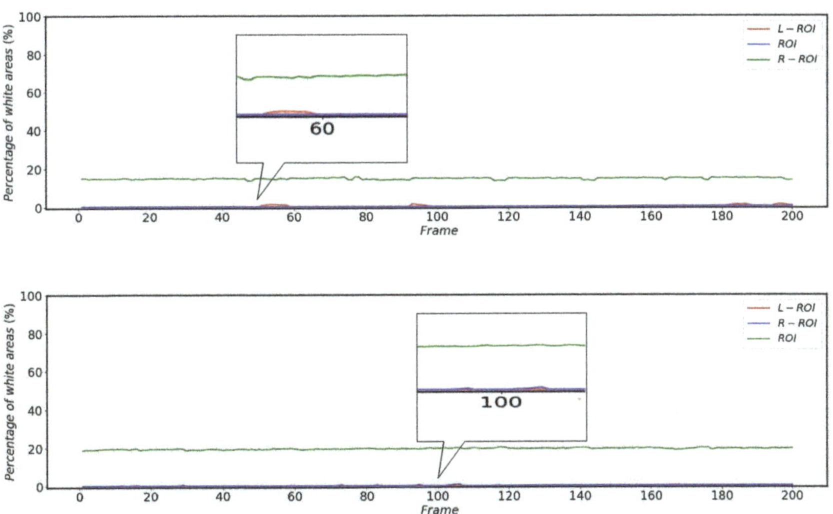

Figure 21. Binary image of different regions and the corresponding vertical projections after the opening operation. (**a**) L-ROI. (**b**) ROI. (**c**) R-ROI.

As shown in Figure 22, the percentage of white areas in the vertical projections of L-ROI and R-ROI is low when the HSR is operating normally without external disturbance, while there is a large percentage of white areas in the vertical projections corresponding to ROI.

Figure 22. Change in the percentage of white areas in the vertical projection of different areas of HSR-A (**top**) and HSR-B (**bottom**) when the HSR is operated without external disturbances.

The impact of the catenary support device on the pantograph detection is much smaller compared to other complex backgrounds, but the percentage of white areas in the vertical projection still reflects the changes brought about by this scenario very accurately. The changes in the percentage of white areas in the vertical projection after different areas in the L-ROI, ROI and R-ROI are affected by the catenary support devices during the operation of the HSR are shown in Figure 23.

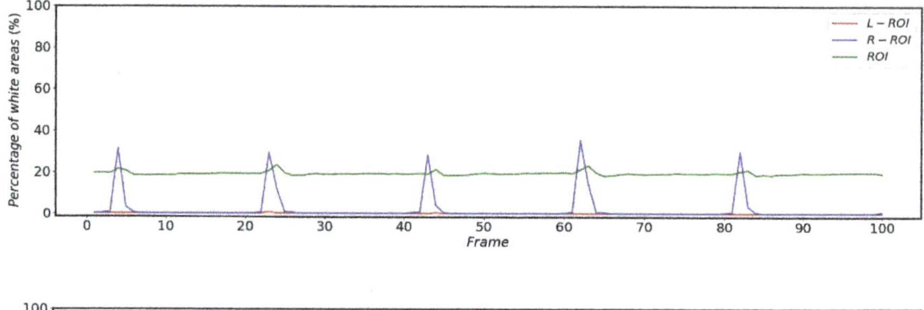

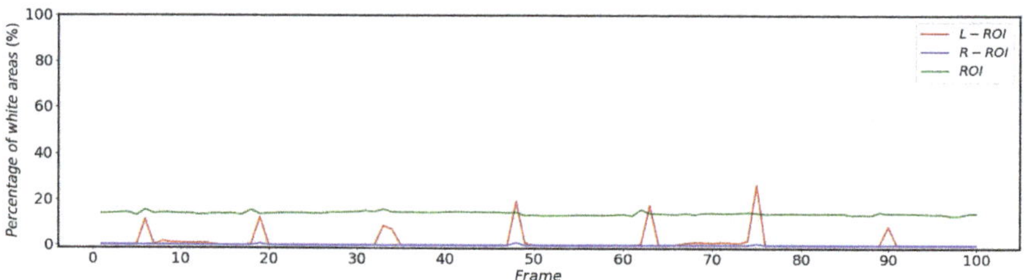

Figure 23. Changes in the percentage of white areas in the vertical projections of different areas of HSR-A (**top**) and HSR-B (**bottom**) during HSR operation after being affected by the catenary support devices.

The effect of bridges on the percentage of white areas in the vertical projection of different regions during HSR operation is shown in Figure 24. Since the HSC angles of HSR-A and HSR-B are different, the bridges do not have the same effect on the percentage of white in the vertical projection areas of L-ROI and R-ROI, but both cause at least one of the L-ROI or R-ROI to have a huge change in the percentage of white area in the vertical projection.

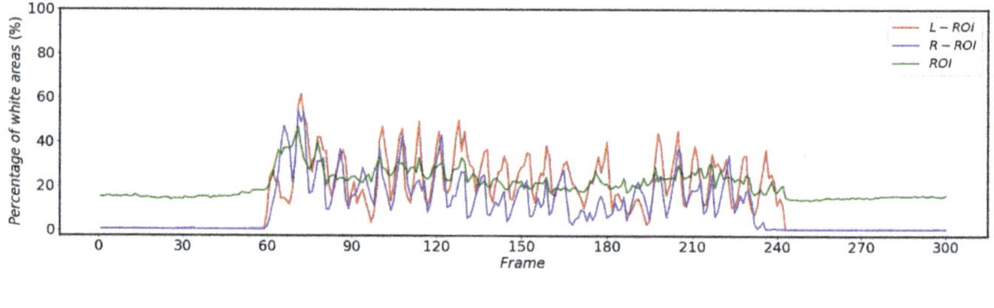

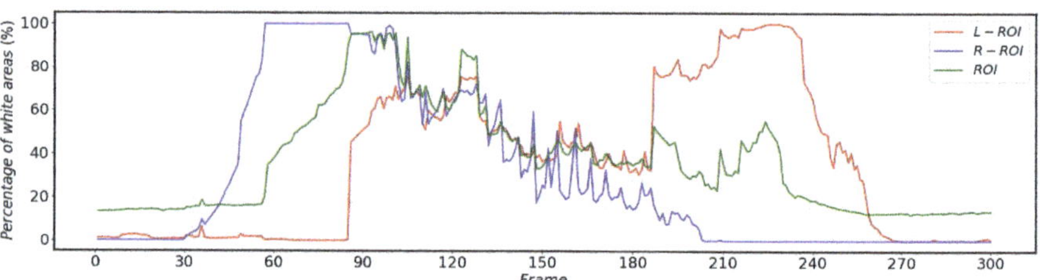

Figure 24. Changes in the percentage of white areas in the vertical projections of different areas of HSR-A (**top**) and HSR-B (**bottom**) during HSR operation after being influenced by the bridge.

The effect of the platform on the percentage of white areas in the vertical projection of the different areas is shown in Figure 25. Furthermore, due to the HSC angle, the impact of the platform on HSR-A and HSR-B is different, but both have an impact on at least one of the R-ROI or L-ROI.

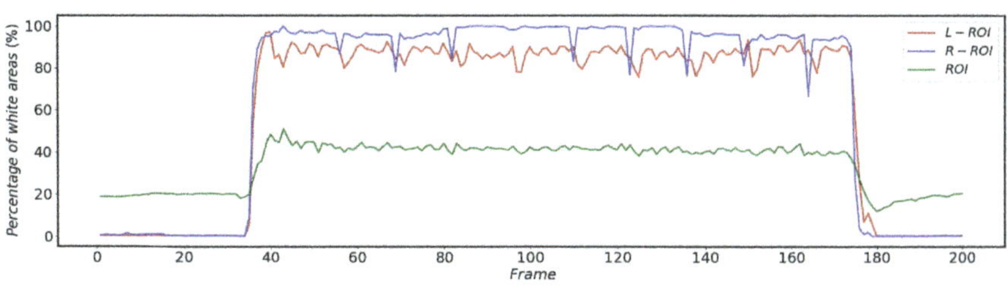

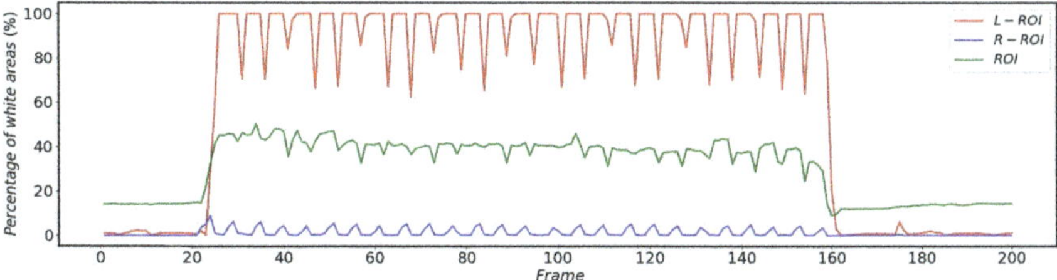

Figure 25. Changes in the percentage of white areas in the vertical projections of different areas of HSR-A (**top**) and HSR-B (**bottom**) during HSR operation after being influenced by the platform.

From Figures 22–25, it can be seen that the percentage of white area in the projection corresponding to ROI does not change much when subjected to complex background interference, while the changes of L-ROI and R-ROI are very obvious after subjected to complex background interference, so this paper mainly detects the presence of complex background interference by the projection of L-ROI and R-ROI areas.

4.5. Overall Process of HSR Complex Background Detection Algorithm

The overall process of the complex background detection algorithm is shown in Figure 26. For a pantograph image captured by a HSC, when it cannot be detected or is detected as abnormal, the complex background detection algorithm is needed to assess whether the current detection result has the possibility of being affected by the complex background.

The specific process is as follows: First, the change of the average grayscale of the current image as a whole and the average grayscale of the previous frame as a whole is used to evaluate whether the detection result may be affected by the drastic change of light before and after the HSR enters and leaves the tunnel. If not, the relationship between the overall average grayscale of the image and the average grayscale of the ROI is used to assess whether the sun may have intruded into the pantograph region and thus influenced the pantograph detection. If the influence of the sun can still be excluded, the detection of the catenary support devices, platforms, and bridges is achieved by vertical projection to finally determine whether the pantograph detection results are influenced by the complex background at this time.

If the influence of complex background on the detection result is excluded by HSR complex background detection algorithm, then there are still two possibilities for the pantograph not to be detected or detected as abnormal: (1) although the current image is not disturbed by complex background, it may be disturbed by other interference which leads to misjudgment of the pantograph, (2) the pantograph does appear abnormal. In this case, the overall algorithm proposed in Section 5.1 of this study is combined to achieve accurate detection of the real situation of pantographs.

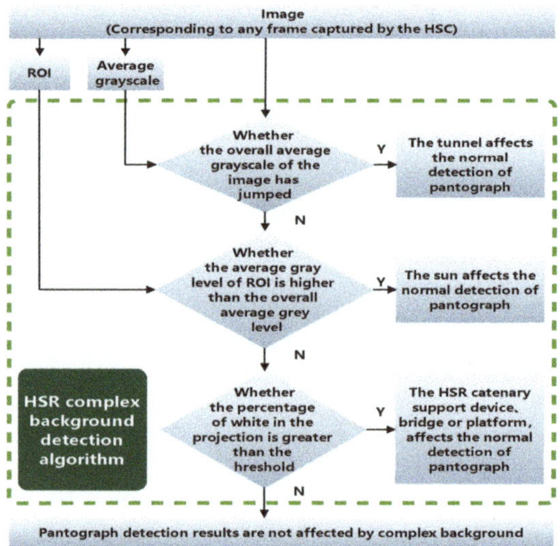

Figure 26. HSR complex background detection algorithm process flow chart.

5. Experiments and Conclusions

5.1. The Overall Process of Pantograph Detection Algorithm

The overall process of the algorithm is shown in the Figure 27, when YOLO V4 cannot detect the pantograph in a frame or detect it as abnormal, the algorithm gives priority to detecting it through the HSC blur and dirt detection algorithm, and when the detection abnormality is ruled out as a result of dirty or blurred screen, then the HSR complex background detection algorithm to determine whether the detection of abnormalities is caused by complex background. Finally, we can realize the accurate judgment of the pantograph state.

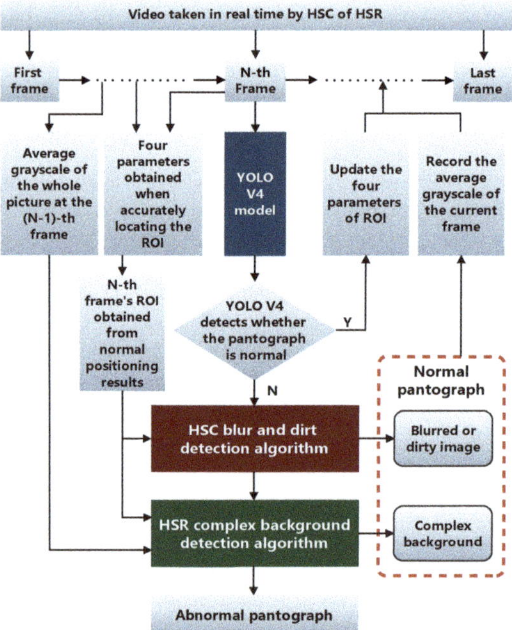

Figure 27. pantograph detection algorithm process flow chart.

5.2. Performance Evaluation of Algorithms under Complex Background Interference

The operation of HSR requires frequent face to the interference and influence brought by scenarios such as catenary support devices, sun, bridges, platforms, and tunnels to pantograph detection. The performance of different methods in detecting pantographs in complex backgrounds is shown in Table 1.

Table 1. Performance of different algorithms when dealing with complex backgrounds.

Method	TM [38]	MS + SIFT [39]	MS + KF [40]	PDDNet [12]	SED [17]	Improved Faster R-CNN [18]	The Method of This Study
Whether the pantograph can be detected correctly under the complex background	×	×	×	×	×	×	✓

Refs. [12,17,18,38–40] all proposed good methods and ideas in order to improve the performance of their respective algorithms in complex backgrounds. However, in the face of more complex background disturbances and effects during the actual operation of HSR, the relevant algorithms still cannot achieve correct detection of pantographs under these complex backgrounds. In contrast, the HSR complex background detection algorithm proposed in this study can well achieve the correct detection and evaluation of the pantograph state under the relevant scenes. The results in Table 1 show that the method proposed in this study is more suitable for the real situation and practical needs of HSR, and performs better under the influence of complex background.

5.3. EOR-Brenner Evaluates the Sharpness of Pantograph Images Captured by HSC

Figure 28 shows the scores of EOR-Brenner on the sharpness of the images captured by two different models under different conditions. Where Frame 1–Frame 100 corresponds to the images captured by HSC during normal operation without any disturbance, Frame 101–Frame 200 corresponds to the blurred image caused by rain affecting the HSC, and Frame 201–Frame 300 is the dirty HSC lens.

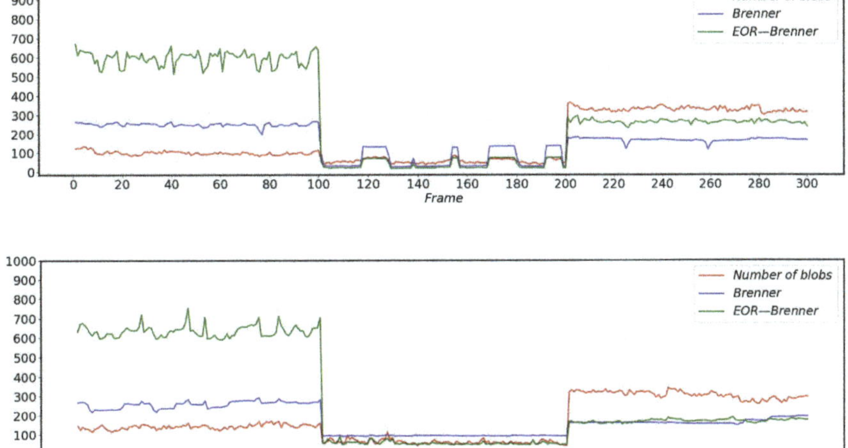

Figure 28. EOR-Brenner evaluation results of images captured by HSR-A and HSR-B under different conditions.

Comparing Figure 28, it can be seen that EOR-Brenner gives higher scores than Brenner for clear pantograph images; for blurred pantograph images EOR-Brenner gives lower scores than Brenner for image sharpness; and the scores are very close when dirty. At the same time, EOR-Brenner has higher distinguishability between clear, blurred and dirty images, while the scores of the original Brenner images are very similar when they are dirty and clear. The improved EOR-Brenner algorithm is more in line with the real operating environment of HSR and better meets the actual needs of HSR operation.

5.4. Evaluation of the Overall Performance of the Algorithm in This Study

The combined test results for complex scenes and blurred and dirty cases are shown in Tables 2 and 3. The red part corresponds to a clear image without interference, the gray part corresponds to a blurred image, the purple part corresponds to an image affected by dirt, and the pink part corresponds to an image disturbed by a complex environment.

Table 2. Comprehensive evaluation of the images presented in this article I.

Image Serial Number	Different Sharpness Evaluation Algorithms								
	Tenengard [41]	Laplacian [42]	SMD [43]	SMD2 [44]	EG [45]	EAV [46]	NRSS [47]	Brenner [33]	EOR-Brenner
Figure 2 left	22.5	4.24	1.81	2.01	9.34	38.18	0.79	252	704
Figure 2 right	31.1	8.25	3.23	5.18	17.26	48.25	0.91	400	876
Figure 5a	9.4	2.18	0.76	0.57	2.31	23.44	0.75	95	55
Figure 5b	10.57	2.49	0.86	0.64	2.46	27.89	0.75	117	64
Figure 6a	31.64	4.45	2.72	2.35	13.92	39.01	0.82	158	228
Figure 6b	32.81	5.52	2.77	2.75	16.32	50.55	0.84	286	476
Figure 10a	26.27	4.55	2.13	2.48	11.98	44.48	0.77	269	686
Figure 10b	39.79	6.76	3.54	5.13	21.42	66.29	0.81	363	767
Figure 11a	24.00	4.56	2.20	2.71	13.62	51.25	0.81	143	310
Figure 11b	14.00	2.54	1.22	1.42	6.77	42.21	0.78	75	285
Figure 12a	42.92	6.78	3.47	3.96	21.19	56.17	0.79	358	613
Figure 12b	31.82	4.84	2.67	3.61	17.03	55.23	0.78	221	346
Figure 13a	27.18	4.12	2.30	2.75	13.49	46.28	0.76	162	356
Figure 13b	10.44	2.21	0.86	0.85	2.43	9.76	0.74	229	230
Figure 13c	20.96	3.70	1.80	1.54	7.97	32.38	0.75	209	342
Figure 13d	10.65	2.34	0.88	0.74	2.38	10.11	0.75	245	246
Figure 14a	46.62	7.53	4.05	6.12	26.28	80.26	0.78	305	924
Figure 14b	39.25	6.14	3.38	3.21	22.02	86.59	0.78	310	551

Table 3. Comprehensive evaluation of the images presented in this article II.

Image Serial Number	Vertical Projection		Average Grayscale		Number of Blob
	L-ROI (%)	R-ROI (%)	Whole	ROI	
Figure 2 left	0.5	0.5	135	146	57
Figure 2 right	0.3	0.4	148	154	62
Figure 5a	0.4	0.4	159	175	30
Figure 5b	0.5	0.3	158	179	29
Figure 6a	3.3	1.1	179	190	481
Figure 6b	6.1	0.7	143	149	445
Figure 10a	1.9	38.6	120	114	61
Figure 10b	14.1	72.0	117	116	73
Figure 11a	3.4	0.5	178	212	69
Figure 11b	0.2	0.5	189	221	44
Figure 12a	46.0	44.7	118	122	140

Table 3. *Cont.*

Image Serial Number	Vertical Projection		Average Grayscale		Number of Blob
	L-ROI (%)	R-ROI (%)	Whole	ROI	
Figure 12b	83.2	67.7	106	100	91
Figure 13a	47.8	69.0	149	154	117
Figure 13b	0	0	2	0	26
Figure 13c	0.5	0.5	52	55	61
Figure 13d	0.5	0.5	250	252	45
Figure 14a	94.3	99.6	112	118	130
Figure 14b	100	7.9	127	141	106

Figure 29 shows the scene of the same HSR running at different times on the same line. Due to the intermittent heavy rainfall, the blurring of the images caused by the HSC affected by rain at different moments is not the same. For the same train on the same line when it is affected differently the results of the clarity algorithm for it are shown in Table 4.

(a) (b) (c) (d) (e) (f)

Figure 29. Scenes taken at different moments of the same HSR in rainy weather. (**a**) Case I. (**b**) Case II. (**c**) Case III. (**d**) Case IV. (**e**) Case V. (**f**) Case VI.

Table 4. Performance of the same HSR at different times with different levels of disturbance.

Image Serial Number	The Actual Time Corresponding to the Scene	Different Sharpness Evaluation Algorithms								
		Tenengard [41]	Laplacian [42]	SMD [43]	SMD2 [44]	EG [45]	EAV [46]	NRSS [47]	Brenner [33]	EOR-Brenner
Figure 29a	16:49:36	16.30	3.15	1.31	1.09	5.69	32.18	0.77	124	149
Figure 29b	16:51:45	9.16	2.45	0.74	0.54	2.20	28.28	0.74	125	63
Figure 29c	18:59:35	22.53	4.72	1.79	1.73	7.98	46.70	0.78	256	756
Figure 29d	19:22:54	23.29	4.82	1.90	1.93	9.12	40.97	0.79	235	764
Figure 29e	20:57:08	9.46	1.76	0.82	0.69	3.45	29.17	0.76	50	81
Figure 29f	22:41:23	9.94	2.37	0.85	0.62	2.54	32.92	0.74	112	59

As can be seen from Tables 2–4, regardless of the cases in which different complex backgrounds or external disturbances affect the pantograph detection of different HSR, or the cases in which the same HSR affects the pantograph detection at different moments due to changes in the external environment, the EOR-Brenner algorithm proposed in this study can accurately evaluate the sharpness of these pantograph images under the influence of disturbances, and the clearer the image, the higher the score. For the blurred pantograph images, the EOR-Brenner algorithm scores them much lower than the normal pantograph images, so as to achieve an accurate judgment of the blurred situation. However, it should be noted that for the images corresponding to Figure 6 when the HSC lens is dirty, a large number of blobs appear on the lens due to the dirt, which will make the image have more edge details at this time, so the EOR-Brenner does not score the dirty image low. However, the number of blobs on the dirty image is much higher than the pantograph images in other cases, so the number of blobs can achieve accurate detection of dirty images.

For the case of complex background affecting pantograph detection, comparing Tables 2 and 3, we can see that the average gray scale of the whole image (Figure 13) before and after entering and leaving the tunnel will suddenly jump to around 0 or 255, while other disturbances affecting the pantograph will not lead to such a drastic change in gray scale value, through this jump in gray scale value can provide a strong basis for whether the high speed rail is driving into the tunnel, so as to exclude the high speed rail The effect on pantograph detection when entering and leaving the tunnel. When the sun affects the pantograph detection (Figure 11) it causes a large difference between the average grayscale of the ROI and the average grayscale of the whole image, while in other cases the difference between the average grayscale of the pantograph area and the whole image is small. Compared with other disturbances, contact network support devices, bridges, and tunnels, when affecting pantograph detection (Figures 10, 12 and 14), cause the white percentage of the vertical projection of at least one of the L-ROI region and R-ROI region to reach more than 35%, while the percentage of the vertical projection of the L-ROI and R-ROI regions in other scenes basically remains around 1%, with the maximum not exceeding 10%. Accurate detection of these scenes can be achieved by this feature.

The results of the comprehensive test for a variety of scenes at the same time are shown in Table 5. Meanwhile, we demonstrate the effectiveness of each module by the ablation experiments shown in Table 6. It is easy to find that the HSR complex background detection algorithm and HSC blur and dirt detection algorithm proposed in this study can greatly improve the accuracy of pantograph inspection evaluation when complex background and external disturbance exist. In general, the algorithm proposed in this study is in line with the real situation of HSR operation and meets the actual needs of HSR operation, which has a greater practical application value.

Table 5. Overall algorithm testing.

Serial Number	Type of Sample	Number of Samples	Total Algorithm Run Time	FPS	Precision
I	Complex backgrounds only	14,985	304 s	49	99.92%
II	Complex backgrounds + Blur	14,999	346 s	43	99.90%
III	Complex backgrounds + Dirt	14,974	349 s	43	99.98%

Table 6. Impact of different modules on the overall algorithm.

	Precision-I	Precision-II	Precision-III
The complete algorithm proposed in this study	99.92%	99.90%	99.98%
− HSR complex background detection algorithm	73.97%	84.76%	85.32%
− HSC blur and dirt detection algorithm	96.24%	73.16%	77.13%
− HSR complex background detection algorithm and HSC blur and dirt detection algorithm	70.36%	57.42%	63.10%

6. Conclusions

The pantograph detection algorithm proposed in this study fully considers the actual needs of HSR operation, and at the same time conducts a comprehensive and synthesize analysis of the complex scenarios and external disturbances that need to be faced during HSR operation. The proposed algorithm achieves precision of 99.92%, 99.90% and 99.98% on different test samples. At the same time, for three different samples, the processing speed of the algorithm per second reaches 49 FPS, 43 FPS and 43 FPS respectively, which meets the requirement of the algorithm to process at least 25 images per second in the actual operation of HSR. This method solves two major difficulties when using neural network to realize pantograph detection: firstly, the current pantograph detection method is easily affected by external interference, and cannot detect and eliminate external interference. Secondly, because the pantograph samples in complex situations are few and difficult to collect, the sample set for training the neural network cannot cover all situations, so the detection accuracy in complex situations is low.

Author Contributions: Methodology, P.T. and Z.C.; Supervision, P.T., X.L., J.D., J.M. and Y.F.; Visualization, Z.C., W.L. and C.H.; Writing—original draft, Z.C.; Writing—review & editing, P.T. and Z.C. All authors have read and agreed to the published version of the manuscript.

Funding: Project supported by the National Natural Science Foundation of China (Nos. 51577166, 51637009, 51677171 and 51827810), the National Key R&D Program (No. 2018YFB0606000), and the China Scholarship Council (No. 201708330502), Shuohuang Railway Development Limited Liability Company (SHTL-2020-13). State Key Laboratory of Industrial Control Technology (ICT2022B29).

Institutional Review Board Statement: Not applicable.

Informed Consent Statement: Not applicable.

Data Availability Statement: Not applicable.

References

1. Tan, P.; Ma, J.e.; Zhou, J.; Fang, Y.t. Sustainability development strategy of China's high speed rail. *J. Zhejiang Univ. Sci. A* **2016**, *17*, 923–932. [CrossRef]
2. Tan, P.; Li, X.; Wu, Z.; Ding, J.; Ma, J.; Chen, Y.; Fang, Y.; Ning, Y. Multialgorithm fusion image processing for high speed railway dropper failure–defect detection. *IEEE Trans. Syst. Man Cybern. Syst.* **2019**, *51*, 4466–4478. [CrossRef]
3. Tan, P.; Li, X.F.; Xu, J.M.; Ma, J.E.; Wang, F.J.; Ding, J.; Fang, Y.T.; Ning, Y. Catenary insulator defect detection based on contour features and gray similarity matching. *J. Zhejiang Univ. Sci. A* **2020**, *21*, 64–73. [CrossRef]
4. Gao, S.; Liu, Z.; Yu, L. Detection and monitoring system of the pantograph-catenary in high-speed railway (6C). In Proceedings of the 2017 7th International Conference on Power Electronics Systems and Applications-Smart Mobility, Power Transfer & Security (PESA), Hong Kong, China, 12–14 December 2017; IEEE: Piscataway, NJ, USA, 2017; pp. 1–7.
5. Gao, S. Automatic detection and monitoring system of pantograph–catenary in China's high-speed railways. *IEEE Trans. Instrum. Meas.* **2020**, *70*, 1–12. [CrossRef]
6. He, D.; Chen, J.; Liu, W.; Zou, Z.; Yao, X.; He, G. Online Images Detection for Pantograph Slide Abrasion. In Proceedings of the 2020 IEEE 20th International Conference on Communication Technology (ICCT), Nanning, China, 28–31 October 2020; IEEE: Piscataway, NJ, USA, 2020; pp. 1365–1371.
7. Ma, L.; Wang, Z.y.; Gao, X.r.; Wang, L.; Yang, K. Edge detection on pantograph slide image. In Proceedings of the 2009 2nd International Congress on Image and Signal Processing, Tianjin, China, 17–19 October 2009; IEEE: Piscataway, NJ, USA, 2009; pp. 1–3.
8. Li, H. Research on fault detection algorithm of pantograph based on edge computing image processing. *IEEE Access* **2020**, *8*, 84652–84659. [CrossRef]
9. Huang, S.; Zhai, Y.; Zhang, M.; Hou, X. Arc detection and recognition in pantograph–catenary system based on convolutional neural network. *Inf. Sci.* **2019**, *501*, 363–376. [CrossRef]
10. Jiang, S.; Wei, X.; Yang, Z. Defect detection of pantograph slider based on improved Faster R-CNN. In Proceedings of the 2019 Chinese Control And Decision Conference (CCDC), Nanchang, China, 3–5 June 2019; pp. 5278–5283.
11. Jiao, Z.; Ma, C.; Lin, C.; Nie, X.; Qing, A. Real-time detection of pantograph using improved CenterNet. In Proceedings of the 2021 IEEE 16th Conference on Industrial Electronics and Applications (ICIEA), Chengdu, China, 1–4 August 2021; IEEE: Piscataway, NJ, USA, 2021; pp. 85–89.
12. Wei, X.; Jiang, S.; Li, Y.; Li, C.; Jia, L.; Li, Y. Defect detection of pantograph slide based on deep learning and image processing technology. *IEEE Trans. Intell. Transp. Syst.* **2019**, *21*, 947–958. [CrossRef]
13. Li, D.; Pan, X.; Fu, Z.; Chang, L.; Zhang, G. Real-time accurate deep learning-based edge detection for 3-D pantograph pose status inspection. *IEEE Trans. Instrum. Meas.* **2022**, *71*, 1–12. [CrossRef]
14. Sun, R.; Li, L.; Chen, X.; Wang, J.; Chai, X.; Zheng, S. Unsupervised learning based target localization method for pantograph video. In Proceedings of the 2020 16th International Conference on Computational Intelligence and Security (CIS), Nanning, China, 27–30 November 2020; IEEE: Piscataway, NJ, USA, 2020; pp. 318–323.
15. Na, K.M.; Lee, K.; Shin, S.K.; Kim, H. Detecting deformation on pantograph contact strip of railway vehicle on image processing and deep learning. *Appl. Sci.* **2020**, *10*, 8509. [CrossRef]
16. Huang, Z.; Chen, L.; Zhang, Y.; Yu, Z.; Fang, H.; Zhang, T. Robust contact-point detection from pantograph-catenary infrared images by employing horizontal-vertical enhancement operator. *Infrared Phys. Technol.* **2019**, *101*, 146–155. [CrossRef]
17. Lu, S.; Liu, Z.; Chen, Y.; Gao, Y. A novel subpixel edge detection method of pantograph slide in complicated surroundings. *IEEE Trans. Ind. Electron.* **2021**, *69*, 3172–3182. [CrossRef]
18. Luo, Y.; Yang, Q.; Liu, S. Novel vision-based abnormal behavior localization of pantograph-catenary for high-speed trains. *IEEE Access* **2019**, *7*, 180935–180946. [CrossRef]
19. Bochkovskiy, A.; Wang, C.Y.; Liao, H.Y.M. Yolov4: Optimal speed and accuracy of object detection. *arXiv* **2020**, arXiv:2004.10934.
20. Girshick, R. Fast r-cnn. In Proceedings of the IEEE International Conference on Computer Vision, Santiago, Chile, 7–13 December 2015; pp. 1440–1448.
21. Ren, S.; He, K.; Girshick, R.; Sun, J. Faster r-cnn: Towards real-time object detection with region proposal networks. *Adv. Neural Inf. Process. Syst.* **2015**, *28*. [CrossRef]

22. Liu, W.; Anguelov, D.; Erhan, D.; Szegedy, C.; Reed, S.; Fu, C.Y.; Berg, A.C. Ssd: Single shot multibox detector. In Proceedings of the European Conference on Computer Vision, Amsterdam, The Netherlands, 11–14 October 2016; Springer: Cham, Switzerland, 2016; pp. 21–37.

23. He, K.; Zhang, X.; Ren, S.; Sun, J. Deep residual learning for image recognition. In Proceedings of the IEEE Conference on Computer Vision and Pattern Recognition, Las Vegas, NV, USA, 26 June–1 July 2016; pp. 770–778.

24. Redmon, J.; Divvala, S.; Girshick, R.; Farhadi, A. You only look once: Unified, real-time object detection. In Proceedings of the IEEE Conference on Computer Vision and Pattern Recognition, Las Vegas, NV, USA, 26 June–1 July 2016; pp. 779–788.

25. Redmon, J.; Farhadi, A. YOLO9000: Better, faster, stronger. In Proceedings of the IEEE Conference on Computer Vision and Pattern Recognition, Honolulu, HI, USA, 21–26 July 2017; pp. 7263–7271.

26. Ju, M.; Luo, H.; Wang, Z.; Hui, B.; Chang, Z. The application of improved YOLO V3 in multi-scale target detection. *Appl. Sci.* **2019**, *9*, 3775. [CrossRef]

27. Kim, S.W.; Kook, H.K.; Sun, J.Y.; Kang, M.C.; Ko, S.J. Parallel feature pyramid network for object detection. In Proceedings of the European Conference on Computer Vision (ECCV), Munich, Germany, 8–14 September 2018; pp. 234–250.

28. Wang, T.; Anwer, R.M.; Cholakkal, H.; Khan, F.S.; Pang, Y.; Shao, L. Learning rich features at high-speed for single-shot object detection. In Proceedings of the IEEE/CVF International Conference on Computer Vision, Seoul, Korea, 27 October–2 November 2019; pp. 1971–1980.

29. Chao, P.; Kao, C.Y.; Ruan, Y.S.; Huang, C.H.; Lin, Y.L. Hardnet: A low memory traffic network. In Proceedings of the IEEE/CVF International Conference on Computer Vision, Seoul, Korea, 27 October–2 November 2019; pp. 3552–3561.

30. Zhang, S.; Wen, L.; Bian, X.; Lei, Z.; Li, S.Z. Single-shot refinement neural network for object detection. In Proceedings of the IEEE Conference on Computer Vision and Pattern Recognition, Salt Lake City, UT, USA, 18–22 June 2018; pp. 4203–4212.

31. Zhao, Q.; Sheng, T.; Wang, Y.; Tang, Z.; Chen, Y.; Cai, L.; Ling, H. M2det: A single-shot object detector based on multi-level feature pyramid network. *Proc. AAAI Conf. Artif. Intell.* **2019**, *33*, 9259–9266. [CrossRef]

32. Liu, H.; Zhang, L.; Xin, S. An Improved Target Detection General Framework Based on Yolov4. In Proceedings of the 2021 IEEE International Conference on Robotics and Biomimetics (ROBIO), Sanya, China, 27–31 December 2021; IEEE: Piscataway, NJ, USA, 2021; pp. 1532–1536.

33. Maier, A.; Niederbrucker, G.; Uhl, A. Measuring image sharpness for a computer vision-based Vickers hardness measurement system. In Proceedings of the Tenth International Conference on Quality Control by Artificial Vision, Saint-Etienne, France, 28–30 June 2011; SPIE: Bellingham, WA, USA, 2011; Volume 8000, pp. 199–208.

34. Kaspers, A. Blob Detection. Master's Thesis, Utrecht University, Utrecht, The Netherlands, 2011 .

35. Zhang, M.; Wu, T.; Beeman, S.C.; Cullen-McEwen, L.; Bertram, J.F.; Charlton, J.R.; Baldelomar, E.; Bennett, K.M. Efficient small blob detection based on local convexity, intensity and shape information. *IEEE Trans. Med. Imaging* **2015**, *35*, 1127–1137. [CrossRef]

36. Bochem, A.; Herpers, R.; Kent, K.B. Hardware acceleration of blob detection for image processing. In Proceedings of the 2010 Third International Conference on Advances in Circuits, Electronics and Micro-Electronics, Venice, Italy, 18–25 July 2010; IEEE: Piscataway, NJ, USA, 2010; pp. 28–33.

37. Xiong, X.; Choi, B.J. Comparative analysis of detection algorithms for corner and blob features in image processing. *Int. J. Fuzzy Log. Intell. Syst.* **2013**, *13*, 284–290. [CrossRef]

38. Thanh, N.D.; Li, W.; Ogunbona, P. An improved template matching method for object detection. In Proceedings of the Asian Conference on Computer Vision, Xi'an, China, 23–27 September 2009; Springer: Berlin/Heidelberg, Germany, 2009; pp. 193–202.

39. Zhou, H.; Yuan, Y.; Shi, C. Object tracking using SIFT features and mean shift. *Comput. Vis. Image Underst.* **2009**, *113*, 345–352. [CrossRef]

40. Li, X.; Zhang, T.; Shen, X.; Sun, J. Object tracking using an adaptive Kalman filter combined with mean shift. *Opt. Eng.* **2010**, *49*, 020503. [CrossRef]

41. Krotkov, E.P. *Active Computer Vision by Cooperative Focus and Stereo*; Springer Science & Business Media: New York, NY, USA, 2012.

42. Riaz, M.; Park, S.; Ahmad, M.B.; Rasheed, W.; Park, J. Generalized laplacian as focus measure. In Proceedings of the International Conference on Computational Science, Krakow, Poland, 23–25 June 2008; Springer: Berlin/Heidelberg, Germany, 2008; pp. 1013–1021.

43. Chern, N.N.K.; Neow, P.A.; Ang, M.H. Practical issues in pixel-based autofocusing for machine vision. In Proceedings of the 2001 ICRA, IEEE International Conference on Robotics and Automation (Cat. No. 01CH37164), Seoul, Korea, 21–26 May 2001; IEEE: Piscataway, NJ, USA, 2001; Volume 3, pp. 2791–2796.

44. Huang, H.; Ge, P. Depth extraction in computational integral imaging based on bilinear interpolation. *Opt. Appl.* **2020**, *50*, 497–509. [CrossRef]

45. Feichtenhofer, C.; Fassold, H.; Schallauer, P. A perceptual image sharpness metric based on local edge gradient analysis. *IEEE Signal Process. Lett.* **2013**, *20*, 379–382. [CrossRef]

46. Zhang, K.; Huang, D.; Zhang, B.; Zhang, D. Improving texture analysis performance in biometrics by adjusting image sharpness. *Pattern Recognit.* **2017**, *66*, 16–25. [CrossRef]

47. Xie, X.P.; Zhou, J.; Wu, Q.Z. No-reference quality index for image blur. *J. Comput. Appl.* **2010**, *30*, 921. [CrossRef]

Article

Deep Learning-Based Intelligent Forklift Cargo Accurate Transfer System

Jie Ren [1], Yusu Pan [2], Pantao Yao [1], Yicheng Hu [1], Wang Gao [3] and Zhenfeng Xue [1,2,*]

[1] Intelligent Perception and Control Center, Huzhou Institute of Zhejiang University, Huzhou 313098, China
[2] Institute of Cyber-Systems and Control, Zhejiang University, Hangzhou 310027, China
[3] Science and Technology on Complex System Control and Intelligent Agent Cooperation Laboratory, Beijing 100191, China
* Correspondence: zfxue0903@zju.edu.cn

Citation: Ren, J.; Pan, Y.; Yao, P.; Hu, Y.; Gao, W.; Xue, Z. Deep Learning-Based Intelligent Forklift Cargo Accurate Transfer System. *Sensors* **2022**, *22*, 8437. https://doi.org/10.3390/s22218437

Academic Editor: Leopoldo Angrisani

Received: 10 October 2022
Accepted: 1 November 2022
Published: 2 November 2022

Publisher's Note: MDPI stays neutral with regard to jurisdictional claims in published maps and institutional affiliations.

Abstract: In this research, we present an intelligent forklift cargo precision transfer system to address the issue of poor pallet docking accuracy and low recognition rate when using current techniques. The technology is primarily used to automatically check if there is any pallet that need to be transported. The intelligent forklift is then sent to the area of the target pallet after being recognized. Images of the pallets are then collected using the forklift's camera, and a deep learning-based recognition algorithm is used to calculate the precise position of the pallets. Finally, the forklift is controlled by a high-precision control algorithm to insert the pallet in the exact location. This system creatively introduces the small target detection into the pallet target recognition system, which greatly improves the recognition rate of the system. The application of Yolov5 into the pallet positional calculation makes the coverage and recognition accuracy of the algorithm improved. In comparison with the prior approach, this system's identification rate and accuracy are substantially higher, and it requires fewer sensors and indications to help with deployment. We have collected a significant amount of real data in order to confirm the system's viability and stability. Among them, the accuracy of pallet docking is evaluated 1000 times, and the inaccuracy is kept to a maximum of 6 mm. The recognition rate of pallet recognition is above 99.5% in 7 days of continuous trials.

Keywords: computer vision and its practical applications; robotics; deep learning; intelligent systems and control theory

1. Introduction

Labor shortage and increasing labor cost are serious problems in today's society. With the concept of Industry 5.0, it is imperative to promote industrial transformation and accelerate the automation and intelligent development of equipment in order to reduce the pressure brought by the rapid rise in labor costs, so more and more intelligent equipment is used in factories and storage environments [1–3]. Nowadays, the status of logistics equipment is increasing, and forklifts, as the main force of logistics handling equipment, have been widely used in many fields, such as factories, ports, and warehouses. However, as the requirements of the operating environment continue to increase, the handling equipment can no longer be operated by human hands, especially in special environments, such as high temperature, and hazardous and explosive environments. Along with the development of driverless technology, forklifts are also slowly approaching advanced technologies, such as intelligent identification, wireless transmission, and autonomous navigation and positioning. Intelligent forklifts can enhance the compound ability of forklifts, improve the overall operation level of forklifts, and gradually add more added value. Therefore, intelligent forklifts are the main development direction of forklifts in the future [2]. The operation of an intelligent forklift is quite straightforward; typically, it inserts and picks up pallets at one preset area before travelling to another to dump them off, accomplishing a full pallet transfer procedure. However, implementing such a straightforward approach presents numerous specific difficulties:

1. Existing methods are more costly for determining whether a pallet is available at a certain location, while the recognition rate is low and also susceptible to interference by environmental factors [4].
2. When the intelligent forklift inserts the pallet, there are problems of high implementation cost [5], low algorithm robustness and insufficient accuracy for the calculation of the relative position between the pallet and the forklift.
3. When controlling the intelligent forklift to insert the pallet after the accurate position is calculated, a fixed control amount is usually used without considering the vehicle running state, which makes the control process deviate and eventually leads to errors in the inserting results [6].

We propose a deep learning-based intelligent forklift accurate cargo transfer system to address the aforementioned issues, as well as to increase the resilience and accuracy of the system. The system consists of various components with various sensors that cooperate to finish the pallet transfer operation. We specifically use RGB surveillance cameras to check whether there is any pallet that need to be transported at the pallet storage location. Once we determine that there are pallets, we send intelligent forklifts to the area. We then use the RGB-D (depth) camera that comes with the intelligent forklift to calculate the precise position of the pallets relative to the forklift. Finally, we use a high-precision control algorithm to control the forklift. The following three aspects make up the majority of the system features:

1. To precisely determine whether there are pallets to be transported in the pallet storage area, we employ a Yolov5-based [7] pallet monitoring system and small target detection module, and its accuracy rate reaches more than 99.5%.
2. To ensure that the final pallet insertion accuracy is within 6 mm, we calculate the real-time pallet position in relation to the intelligent forklift using the pallet position recognition system based on 3D Hough network [8].
3. We present a high-precision tracing control approach for intelligent forklifts in order to increase the control accuracy, and the docking results obtained from 1000 experiments have an error of no more than 6 mm.

In our warehouse, we employ cameras to monitor pallets and intelligent forklifts to insert and remove pallets, as shown in Figure 1.

(a) Pallet monitoring situation   (b) Pallet insertion

Figure 1. System operation diagram. We have constructed the entire system that is detailed in this paper in the warehouse. One of the eight cameras in the pallet monitoring system, which can monitor the presence of pallets in the storage area and mark them with red boxes when they are found, is illustrated in (**a**). The intelligent forklift arrives at point (**b**), determines the location of the pallets, and then executes the insertion and extraction procedure depicted in the figure after realizing that the pallets need to be moved.

In the rest of the paper, we discuss related work in Section 2, describe our system in Section 3, then present experimental results in Section 4, and give conclusions in Section 5.

2. Related Works and Background

In this section, we discuss the work related to pallet monitoring, pallet position recognition, and intelligent forklift control.

2.1. Pallet Monitoring

When intelligent forklifts are first put into use, it is usually the human who determines whether there are pallets to be inserted and picked up at a specific point. Workers usually hold devices, such as tablet computers or pagers, and send commands to the intelligent forklifts to insert and pick up pallets. However, this approach necessitates human involvement, which wastes labor and is ineffective.

In some projects, the use of sensor-assisted automatic identification techniques has begun [9]. In such projects, sensors are typically installed at pallet storage locations to detect the presence or absence of pallets [10], and the results are then transmitted to the dispatching system via a network cable so that the intelligent forklifts can determine whether a pallet needs to be moved at a specific location. However, using this approach in a large-scale storage environment requires the deployment of sensors at each pallet storage location, which greatly increases the difficulty and cost of implementation.

Recent studies have looked into using RGB surveillance cameras to detect the presence or absence of pallets [11]. This recognition is typically based on the conventional image recognition scheme, which assumes that the appearance of the pallet is mostly visible and clearly distinguished from its surroundings, and then extracts the pallet from the image using techniques such as image segmentation, template matching, etc. This method does not account for the fact that a pallet may have goods covering all or nearly all of its surface, which makes it easy to mistake a pallet for nothing because the camera cannot gather enough data on the pallet's color and contours.

In 2021, Joo et al. proposed a Yolov3-based pallet recognition method [12] which is designed for the industry and can recognize pallets more steadily than conventional image recognition techniques. However, the technique necessitates that the camera be placed in close proximity to the pallet in order to collect data. One camera cannot effectively monitor a large area of pallets, and the recognition rate is low for pallets with a small pixel share.

2.2. Pallet Position Recognition

Pallet position recognition relative to forklift has always been an industry challenge. The pallet must frequently be placed manually or mechanically at the exact location (error less than 1 cm) on the shelf during the initial stages of unmanned forklift use. The intelligent forklift only needs to get to the fixed position each time to finish inserting and retrieving the pallet in this situation because the position of the pallet and the shelf is essentially fixed. This method is not suitable for the automated operation of the plant because it requires too much accuracy in pallet placement, and if there is a mistake, it is easy to happen that the forklift cannot insert the pallet and needs manual assistance.

A technique for using auxiliary markers, such as QR codes, for position recognition has surfaced in the industry as a solution to such issues [13]. The pallets are marked with additional markers, and an on-board scanning gun is used to find the markers. Because it can calculate the position of the pallet in relation to the forklift based on the location of the QR code while entering the pallet information, it has been widely used in the industry. However, we prefer a method that identifies pallets based on their own shape, texture, and other information rather than methods that require markings to be posted on each pallet, which requires a lot of work in the pre-deployment stage.

Garibotto et al. [14,15] proposed a vision-based algorithm to detect the central hole of a pallet, where the hole features of the pallet are extracted after a pre-segmentation of the image, and then the geometric model of the pallet is projected onto the image plane for position estimation. However, the traditional image approach used by this method to identify the position of the pallet hole makes it more susceptible to interference. This is especially true if the shape of the portion of the goods above the pallet is similar to the shape of the hole position, which is typically a simple rectangle.

2.3. Forklift Precision Control Algorithm

The more common forklift control algorithm is based on the improvement of PID control [16], and the desired control effect is achieved by adjusting different control parameters according to the actual usage. However, due to its ease of use and constrained range of adjustment, this method is difficult to adapt to complex factory or warehouse environments and frequently exhibits the trait of low control accuracy in use.

Jiang, Zhizheng et al. proposed a Robust H-based forklift control algorithm [17] to model the dynamics of an electric power steering (EPS) system of an electric forklift. The standard H control model of the EPS system is transformed to derive the generalized equation of state for the EPS of an electric forklift. The principle of genetic optimized robust control is described, and the constraint function of genetic algorithm (GA) is constructed for the parameter optimization of the weighting function of the H control model, and the genetic optimized robust controller is derived. The accuracy and robustness of the forklift control process are effectively enhanced. This method, however, also only focuses on the function of EPS in forklift control and neglects to design for the actual operating circumstances during forklift operation or take into account the dynamic adjustment of the control volume to increase control accuracy.

3. Intelligent Forklift Cargo Precision Transfer System

In a factory or warehouse, intelligent forklifts are needed to accurately insert and transport pallets, as well as autonomously determine whether there is any pallet in the pallet storage area that need to be transferred. We introduce an intelligent forklift cargo precision transfer system to carry out this function, as depicted in Figure 2. In order to determine whether there is any pallet that need to be transported at the pallet storage location, we use a standard RGB surveillance camera. The forklift is then dispatched to the area of the pallet, and the exact position of the pallet is recognized by the RGB-D camera that comes with the forklift. Finally, according to the recognized exact position, a high-precision control algorithm is used to control the forklift to insert and pick up the pallets.

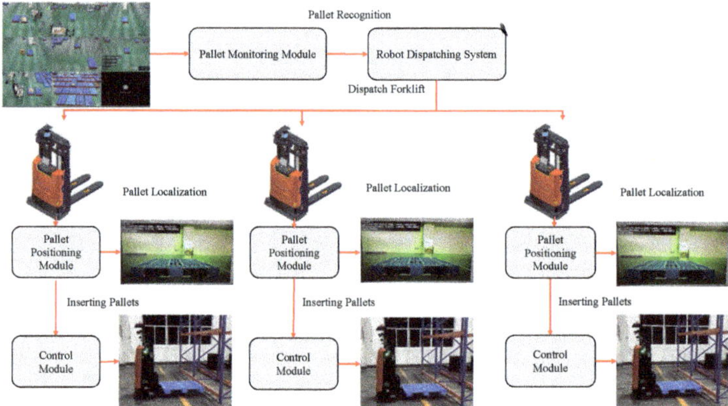

Figure 2. The overall flow of the intelligent forklift precision cargo transfer system. RGB camera captures images of the pallet storage area and transmits them to the pallet monitoring module, which identifies the pallets and informs the dispatching system. The dispatching system dispatches forklifts to insert and pick up the identified pallets. The forklift reaches the pallet and uses the pallet positioning module to identify the pallet position, and after the position is identified, it is handed over to the high precision control module to control the vehicle to insert and pick up the pallet.

3.1. Pallet Monitoring Module

Pallet monitoring is the cornerstone of pallet transfer by forklift, and we carry out this monitoring task using standard RGB monitoring cameras. One advantage of these RGB cameras is that they are less expensive than photoelectric sensors or RGB-D cameras, and they are also much easier to deploy due to the fact that one RGB monitoring camera can cover at least 50 pallet positions. To prevent the influence of items placed on the pallets on the recognition outcomes, we used the side aperture feature of the pallets, which is not easily covered up, as the identification mark. We used a Yolov5-based network for pallet monitoring. However, the size of the pixels occupied by the same pallet in the same image is different because of the viewing angle when the surveillance camera is placed. The traditional Yolov5 scheme will easily fail when the pallet is far from the camera because the pallet's pixels in the field of view will be small. In order to handle these situations, we added a small target detection module and a convolutional block attention module (CBAM).

3.1.1. BackBone Modules

CspDarknet53 [18] is a Backbone structure based on Darknet53 [19], which contains 5 CSP modules. Each CSP module's convolution kernel is 3×3 in size with stride = 2, so it can function as a downsampling. The input image is 608×608, and since Backbone has 5 CSP modules, the pattern of feature map change is $608 \rightarrow 304 \rightarrow 152 \rightarrow 76 \rightarrow 38 \rightarrow 19$. After 5 CSP modules, the 19×19-inch feature map is obtained.

3.1.2. Neck Module

The feature extractor of this network uses a new enhanced bottom-up pathway FPN [20] structure that improves the propagation of low-level features. Each stage of the third pathway takes the feature map of the previous stage as input and processes them with a 3×3 convolutional layer. The output is added to the feature maps of the same stage of the top–down pathway via lateral connections, and these feature maps inform the next stage. Adaptive feature pooling is also used to recover the corrupted information path between each candidate region and all feature levels, aggregating each candidate region at each feature level to avoid arbitrary assignment.

3.1.3. Small Target Detection

We incorporate a transformer prediction head (TPH) into Yolov5 in order to detect small targets. A low-level, high-resolution feature map, which is more sensitive to small objects, is used to generate the added TPH. We also swap out some convolution and CSP blocks with the transformer encoder block. The transformer encoder block in CSPDarknet53 has more information acquisition advantages than the original bottleneck block. The first sub-layer in each transformer encoder block is a multi-headed attention layer, and the second sub-layer (MLP) is a fully connected layer. Each sub-layer is connected by residuals. The transformer encoder block improves the ability to record various local details and can also make use of the self-attentive mechanism to unlock the potential of feature representation.

3.1.4. Convolutional Block Attention Module (CBAM)

CBAM [21] is a simple but effective attention module. It is a lightweight module that can be integrated into a CNN and can be trained in an end-to-end manner. According to the experiments in the paper [22], the performance of the model is greatly improved after integrating CBAM into different models for different classification and detection datasets, which proves the effectiveness of the module. In images of pallet surveillance, large coverage areas can contain a high number of interference terms. Using CBAM can extract attention regions and help the network resist confusing information and focus attention on useful target objects.

3.2. Pallet Positioning Module

The relative poses between the target pallet and the intelligent forklift must be identified after the forklift approaches the target pallet in order to provide precise real-time pose for the forklift's control algorithm. In order to accomplish this, we suggest a novel approach based on a deep 3D Hough voting network. To be more specific, we employ a RGB-D camera to record the color and depth data of the pallet. We then input the color and depth data into a feature extraction module to extract the surface features and geometric information of the pallet. This extracted information is then fed to a key point detection module M_k to predict the offset of each point relative to our specified key points, which are typically defined as the 8 corner points of the two apertures of the pallet. Additionally, we employ a center voting module M_c to predict each point's offset from the target center and an instance segmentation module, M_s, to predict the label of each point. Finally, the obtained 8 key points are used to estimate the pallets' poses relative to the forklift by the least squares method. The whole algorithm flow is shown in Figure 3.

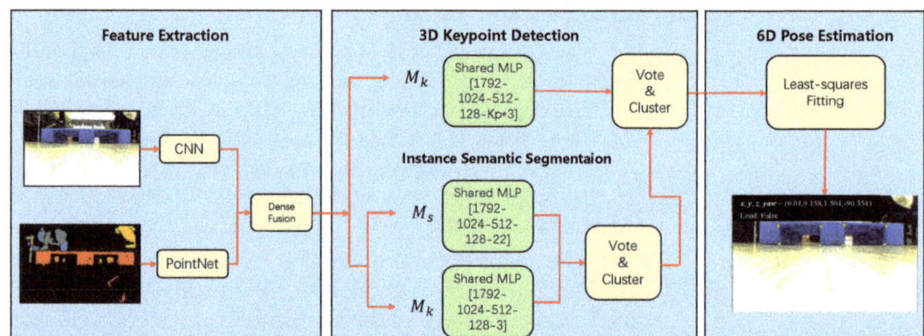

Figure 3. Flow of pallet position recognition algorithm. To predict the translational offsets to the key-points, center points, and labels of each point, the feature extraction module extracts features from the RGB-D images. These features are then fed into the M_k, M_s, and M_c modules. The key-points in the same instances are then voted to the key-points of the instance they belong to after a clustering algorithm is used to distinguish between the various instances. Finally, a least squares method is used to fit the bit pose based on the eight extracted key points. K_p is the number of key points.

3.2.1. Key-Point Detection Module

The pallet image is fed to the key-point detection module M_k after feature extraction to detect key-points on the pallet. The main function of M_k is to predict the Euclidean translation offset from the visible points to the key-points, and these visible points and the predicted translation offset work together to finally be able to vote out the key-points. These voted points are pooled together by a clustering algorithm [23], and the center of the pool is taken as the final key point.

The loss function of M_k is calculated as follows. Given a set of extracted feature points $\{p_i\}_{i=1}^{N}$, where $p_i = [x_i, f_i]$, x_i denotes the 3D coordinates of the points and f_i denotes the point features. Similarly $\{kp_j\}_{j=1}^{M}$ is used to denote the selected key points. We use $\left\{of_i^j\right\}_{j=1}^{M}$ to denote the translation offset of the i-th point with respect to the j-th key point. Thus, the key point can be represented as $kp_i^j = x_i + of_i^j$. To supervise the learning of of_i^j, we use the loss function:

$$L_{\text{key-points}} = \frac{1}{N} \sum_{i=1}^{N} \sum_{j=1}^{M} \left\| of_i^j - of_i^{j*} \right\|$$ (1)

$of_j *(i)$ is the true value of the translation offset. M is the number of selected key-points, which is usually selected as 8 in the system. N is the number of feature points.

3.2.2. Instance Segmentation Module

In order to handle the case where there are multiple objects in the image, i.e., multiple goods or other objects in addition to the pallet, it is necessary to segment the other objects from the pallet in order to extract the key points more accurately. As a result, we present the M_s instance segmentation module. The instance segmentation module M_s predicts the semantic label of each point using the extracted point-by-point data. To supervise the learning of this subject, we employ Focal Loss [24].

$$L_s = -\alpha(1 - q_i)^\gamma \log(q_i), \text{ where } q_i = c_i \cdot l_i \tag{2}$$

with α the α-balance parameter. γ the focusing parameter. c_i the predicted confidence for the i-th point belongs to each class and l_i the one-hot representation of ground true class label.

We also employ a center voting module, M_c, to help distinguish between the various instances. We use a similar module to CenterNet [25], but expand the center points from 2D points to 3D points. This module is able to predict the Euclidean translation offset of each point to its object center in order to achieve a better instance segmentation. Similar to the key-point detection module mentioned above, it can be used to regard the object's center as a certain kind of key-point. We denote by Δx_i the translational offset of each feature point with respect to its object center, then the learning of Δx_i can be supervised by the following loss function.

$$L_c = \frac{1}{N} \sum_{i=1}^{N} \|\Delta x_i - \Delta x_i^*\| \tag{3}$$

where N denotes the total number of seed points on the object surface and $\Delta x^* i$ is the ground truth translation offset from seed p_i to the instance center. It is an indication function indicating whether point p_i belongs to that instance.

3.2.3. Network Architecture

As shown in Figure 3, the first part of the network is a feature extraction module. In this module, a PSPNet [26] is used to extract the appearance information in RGB images. PointNet++ [27] extracts the geometric information in the point cloud and its normal mapping. The two are fused by the DensionFusion block [28] to obtain the combined features of each point. The next M_k, M_s and M_c consist of shared multilayer perceptrons (MLPs). We supervise the learning of module M_k, M_s and M_c jointly with a multi-tasks loss:

$$L_{multi-task} = \lambda_1 L_k + \lambda_2 L_s + \lambda_3 L_c \tag{4}$$

We sample $n = 12{,}288$ points for each RGB-D image frame and set $\lambda_1 = \lambda_2 = \lambda_3 = 1.0$.

3.2.4. Least Squares Fitting

In the least-square fit [29], we denote the final set of 8 key points inferred by the network as $\{kp_i\}_{j=1}^{M}$. The coordinates of this point set are in the camera coordinate system, and the corresponding set of 8 points in the pallet coordinate system is denoted as $\{kp_i'\}_{j=1}^{M}$. To obtain the bit-pose relationship (R, t) between the pallet and the camera, we will minimize the set of these two points between the following loss functions.

$$L_{least-squares} = \sum_{j=1}^{M} \left\| kp_j - \left(R \cdot kp_j' + t \right) \right\|^2 \tag{5}$$

The camera and the intelligent forklift are often fixedly coupled when in use. After obtaining the relative position of the pallet and the camera, only one step of external reference conversion needs to be added to obtain the relative position between the forklift and the pallet.

3.3. High Precision Control Module

Once the pallet position has been determined, we must accurately control the forklift to insert the pallet in that position. Traditional forklift control models [30] and algorithms are overly simplistic, especially the actual running condition of the vehicle and the amount of dynamic adjustment control are not considered enough, which frequently results in significant inaccuracies when used in practice. We propose a high-precision trajectory control method for intelligent forklifts that incorporates forklift motion cycle prediction into the control process, continuously updates the intelligent forklift prediction model, and determines the optimal control amount in order to achieve the goal of increasing control accuracy.

In order to describe the motion of the intelligent forklift in the prediction cycle, a discrete prediction model based on the forklift model is required. Using the forklift motion model as the base model, a prediction model for the non-linear optimization problem is established and a discrete vehicle model in incremental form is performed, as shown:

$$u(k) = u(k - 1) + \Delta u(k) \tag{6}$$

As shown in Figure 4, according to the recurrence relationship of the incremental model, the i-th future prediction state can be represented by the initial state and the sequence of i control increments to complete the construction of the non-linear prediction model.

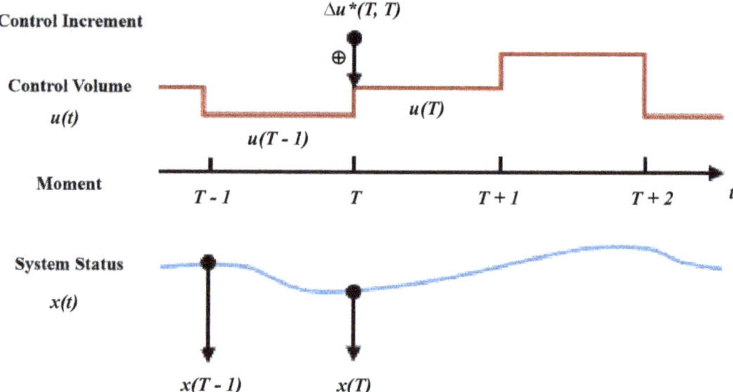

Figure 4. Incremental model recurrence relationship. Get the value at different times.

Combining the path tracking objective function based on the approximate tracking error [22], the non-linear prediction model and constraints [31], the path tracking optimization problem based on the approximate tracking error is constructed as shown:

$$\min J = \sum \left(\|e\|^2 + \|\Delta u\|^2 + \|u\|^2 \right) \tag{7}$$

Additionally, the constraints are added to the optimization problem to obtain the prediction state and control increment by minimizing the sum of squares of approximate tracking error, control increment, and control quantity in the prediction cycle to complete the update of the model and control quantity.

In general, we update the prediction model based on the observed values and the discrete prediction model, establish the path tracking objective function based on the approximation error, obtain the optimal control volume, and combine the previous moment control volume output to the controlled vehicle to complete the high-precision tracking control.

4. Experiment Results

To guarantee that the findings obtained in this article are consistent with the results in actual use, we run the system through real-world scenarios in order to obtain more realistic and accurate results.

4.1. Experiments Environment Construction

The whole system consists of three main parts: pallet monitoring module, pallet positioning module, and high-precision control module. The supporting scheduling system is not discussed in any detail in this work. The forklift body control algorithm, or the high-precision control module, does not need to build an experiment environment and can be tested directly with the forklift.

4.1.1. Pallet Monitoring Module

The monitoring system consists of 8 network RGB cameras and a dispatch server, and 8 network cameras are connected to the server through a switch. We placed 53 pallets in the field of view of camera 8-th and multiple pallets in different locations and numbers in the field of view of other cameras, while placing goods on the pallets to simulate real scenarios. We expect the algorithm to have a good recognition rate for this complex and variable situation. Figure 5 shows the working situation of the monitoring system.

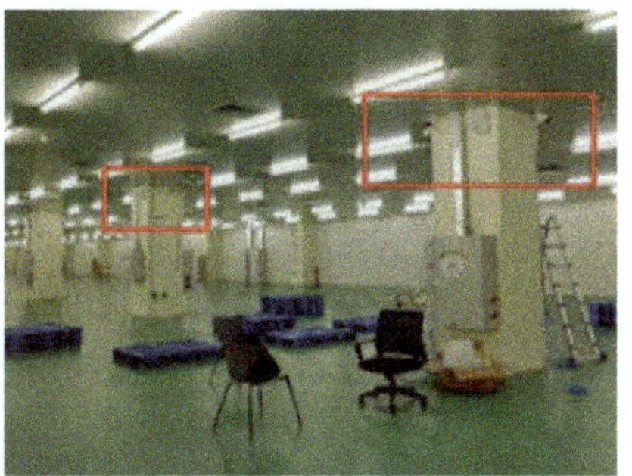

(a) Position of surveillance camera   (b) Installation angle of camera

Figure 5. Pallet monitoring deployment. The surveillance cameras were mounted as high as possible and angled so that more pallets could be monitored. The network camera was installed at a height of approximately 3.27 m and the network camera was overhead (with vertical lines) at an angle of approximately 50 deg.

4.1.2. Pallet Positioning Module

A RGB-D camera was used for pallet position recognition, which was installed in a fixed position on the forklift and the data were uploaded to the forklift mounted IPC for position calculation via a network cable. In order to avoid the fork tines of the forklift from appearing in the image, we adjusted the camera to a suitable position so that it could see the complete pallet hole position without seeing the fork tines of the forklift. Figure 6 shows the RGB-D camera and its recognition results.

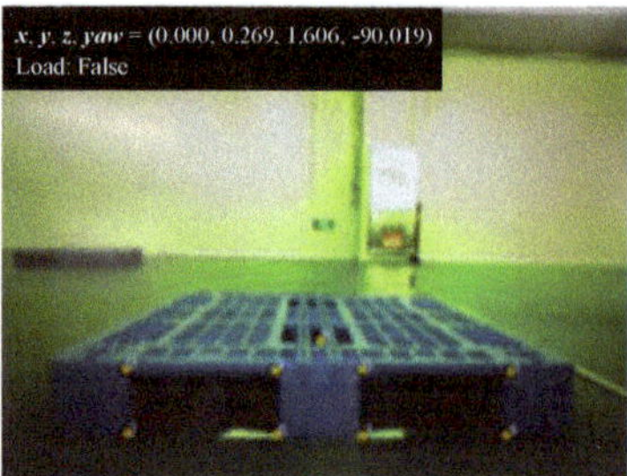

(a) Position of surveillance camera   (b) Installation angle of camera

Figure 6. Pallet positioning module deployment. An on-board RGB-D camera was used and the camera was adjusted to a suitable position. The 8 corner points of the pallet aperture are recognized and the result of the pose calculation is shown in the upper left corner of the image.

4.2. Experiment Results

4.2.1. Pallet Monitoring Module

In order to test the monitoring effect of pallets from multiple cameras, we placed pallets at different locations in the 8-way camera field of view for actual testing. Considering that pallets are likely to be stacked and loaded with goods at customer sites, we placed 2 and 3 layers of pallets in a large number of pallets and placed goods at the top edge of the pallets. We used the side aperture feature of the pallets, which is not easily concealed, as the identification mark in order to prevent the influence of the goods on the pallets on the recognition results. The placement of the pallets under each camera is shown in Figure 7.

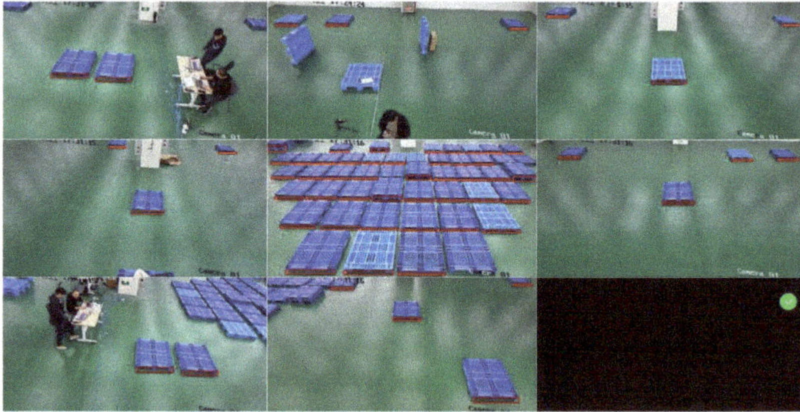

Figure 7. Pallet placement under each camera. Different number of pallets with different number of layers were placed under different cameras and tested repeatedly.

The figure shows how the pallet monitoring system can accurately present the results of pallet recognition for multiple scenarios. Our algorithm places a red box in the side hole position of the recognized pallet to show that a pallet was identified. To see the recognition results more clearly, we use the camera 8-th with the highest number of pallets for analysis.

Before adding the small target detection and CBAM described in 3-A, the recognition effect of camera 8-th is shown in Figure 8. It is obvious that some pallets located further away from the camera cannot be detected, which is typically caused by unreliable recognition brought on by low pixels.

Figure 8. The recognition effect of camera 8-th. The majority of the pallets in the field of view are identified, however certain specific pallets are still not.

We have used small target detection and CBAM to solve such problems, and the recognition effect after addition is shown in Figure 9. It can be clearly seen that the unrecognized pallets in Figure 8 have been recognized in Figure 9, and all unobstructed pallets in the field of view are recognized.

Figure 9. Recognition effect after using small target detection and CBAM. All pallets in the field of view, including multi-layer pallets and cargo pallets, are recognized.

To better reflect the advantages of our algorithm, we list the comparison of existing algorithms with our algorithm in Table 1.

Table 1. Comparison of pallet monitoring algorithms.

Algorithm	Method	Advantages	Disadvantages
Algorithm in [9]	photoelectric sensors	High recognition rate	High cost
Algorithm in [11]	image recognition	Low cost	Easily disturbed
Algorithm in [12]	Deep learning based on Yolov3	High stability	Low recognition rate for small targets
Our algorithm	Deep learning based on Yolov5	Low cost, high recognition rate	Algorithm is complicated

As can be seen from the table, the implementation cost of the photoelectric sensor method is high; the traditional image method is susceptible to interference; and the deep learning method based on Yolov3 does not have a high recognition rate for small targets. Although our method, with low cost, still has a high recognition rate for small targets.

In order to test the reliability of the system operation, we designed the experiment for continuous 7 × 24 h non-stop testing. We adjust the position of multi-layer pallets and goods within the scene before the start of each day's experiment to ensure that the conditions are different each time. Following the change, we will count the unobstructed pallets and enter that number as a parameter in the system. The system will calculate the recognition rate by comparing the pallets it has identified with the input parameter. Table 2 shows the number of pallets and the recognition rate during the 7-day experiments.

Table 2. Pallet monitoring reliability experiments.

Day	Number of Pallets	Recognition Accuracy (%)
1	72	99.82
2	70	99.76
3	72	99.68
4	68	99.86
5	71	99.69
6	74	99.58
7	69	99.71

As can be seen in Table 2, the recognition rate is less than 100%, indicating that there are still cases where individual pallets are not recognized. The recognition rate exceeds 99.5% for about 70 pallets tested continuously for 7 days, including single-layer, multi-layer, with or without goods. For the situation where a large number of pallets appear under one camera, recognition can still be performed stably as long as the pallets are not blocked from each other before (i.e., a reasonable distance between pallets is set). The eight cameras are able to output data streams together for simultaneous recognition of pallets in the area.

4.2.2. Pallet Positioning Module and High Precision Control Module

In the pallet position recognition module, we use RGB-D camera to collect data from several different positions and angles of the pallet for calculation, and provide the results to the intelligent forklift to control the forklift for precise pallet insertion. Since we expect to obtain the overall accuracy of pallet insertion by the forklift, we do not distinguish the positional accuracy from the control accuracy; instead, we judge the ultimate total insertion accuracy.

We collect a lot of data in order to ensure that the findings of our trials covered the majority of usage cases. Given that the relationship between forklifts and pallets is typically not fixed before pallets are inserted, we set up a variety of starting positions: the distances between the pallet hole plane and camera plane are 1 m, 2 m, and 3 m; the horizontal distances are 0 m, −0.2 m, and +0.2 m; and the relative angles between the camera and pallet are 0deg, +15 deg, and −15 deg. Additionally, we established two separate heights, with the pallet placed on the ground being recorded as 0, and the pallet placed on a shelf that is 40 cm high being recorded as 0.4 m. This is because a single pallet may be placed on either the ground or a shelf. We carried out pallet docking approximately 10 times for each scenario, comparing the inaccuracy of each result with the initial result. In other words, we completed a total of 1000 dockings, of which 550 were for the 1-layer pallets, 180 were for the 2-layer pallets, and 260 were for the 3-layer pallets.

We have placed the pallets in different positions and layers and recorded the final error. We recorded the error value of each result compared with the first result and drew it in Figure 10, where the x-axis represents the number of docking times and the y-axis represents the result error value.

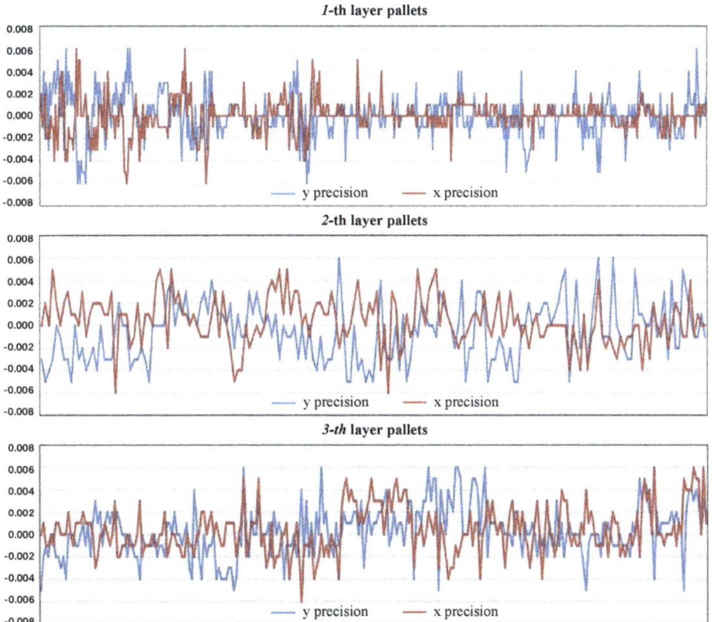

Figure 10. Results of pallet insertion experiment. 1-layer pallets test 550 times, 2-layer pallets test 180 times, 3-layer pallets test 260 times, and the maximum error of 6 mm.

Figure 10 shows that the accuracy in both the x-direction and the y-direction can be controlled to about 5 mm. However, in some circumstances, particularly when dealing with three-layer pallets, the error will increase to 6 mm because multi-layer pallets produce more interference than single-layer pallets.

Overall, our pallet docking accuracy can be controlled within ±6 mm and the coverage range is from 1–3 m and from −15° to +15°. When compared to the measures in the literature [32], the coverage range is roughly the same, but it does not test for severe pallet angle deflection. More importantly, the pallet accuracy in the literature [32] is only ±3 cm, which is much lower than our ±6 mm accuracy. Similar coverage and coverage angle experiments to ours were carried out in the literature [33], but their maximum error of recognition was 10.5 mm after only 135 experiments, which was 75% higher than our maximum error of 6 mm. Its mean error and standard deviation are also significantly higher than ours. Meanwhile, our number of experiments was 7.4 times higher than his, and we measured the final error after pallet docking, which is the accumulation of recognition error and control error values, indicating that our identification error values are lower. On the other hand, the literature [33] used two sensors to achieve the results expressed in the article, while we obtained better results by using only one RGB-D camera. The experiment results are shown in Table 3.

Table 3. The effect comparison of three algorithms.

Algorithm	Range	Max Error (mm)	Mean Error (mm)	Std Error (mm)
Algorithm in [32]	1.5–4.5 m	30	n/a	n/a
Algorithm in [33]	2–4 m, 15°	10.5	7	3.9
Our algorithm	1–3 m, 15°	6	3.69	0.3

The data in the table shows that the maximum error of the existing algorithm for pallet position recognition can be 1–3 cm, which is a risk that the fork tines cannot enter the pallet hole for the high-precision pallet insertion action of the forklift. Our module is able to reduce the maximum error to 6 mm, which fully meets the needs of practical use. At the same time, our system requires only one RGB-D camera to collect data without the assistance of other sensors, which is much less costly.This demonstrates that our approach outperforms the majority of systems now used in industrial settings in terms of accuracy, resilience, and cost.

According to the above experimental results, the recognition rate of our pallet monitoring system has been significantly improved after the addition of small target detection, and the recognition rate reached 99.5% under 7 days of continuous experiments; our position recognition and control algorithms work together to control the forklift interpolation accuracy within 6 mm in 1000 experiments, which fully meets the requirements of accuracy in practical use and outperforms existing algorithms in both coverage and accuracy.

5. Conclusions

We propose an intelligent forklift cargo accurate transfer system, which consists of three main parts: pallet monitoring module, pallet positioning module, and high-precision control module. The system creatively introduces small target detection into pallet recognition to improve the recognition rate; using Yolov5-based pallet pose detection algorithm to improve the coverage and recognition accuracy of the algorithm. For the whole system, we proved its effectiveness and reliability by actually collecting a large amount of data. Among them, the recognition correct rate of pallet monitoring module reaches more than 99.5% in 7 days continuous experiments, and the insertion error of the intelligent forklift is below ±6 mm after 1000 experiments through pallet positioning and control algorithm. The performance of the whole system is greatly higher than the existing common systems, and has been used in factories and warehousing environments on the ground. We will further investigate the use of lightweight networks to refine our system in order to reduce computational resource consumption and computation time. In this way, the cost of commercializing the system can be further reduced.

Author Contributions: Conceptualization, J.R. and Y.P.; methodology, J.R.; software, Y.H.; validation, J.R., P.Y. and Y.H.; formal analysis, Y.P.; writing—original draft preparation, J.R.; writing—review and editing, Y.P. and Z.X.; supervision, W.G. and Z.X.; project administration, Y.H. All authors have read and agreed to the published version of the manuscript.

Funding: This research was funded by Zhejiang Province Science and Technology Plan Project (2022C01214).

Institutional Review Board Statement: Not applicable.

Informed Consent Statement: Not applicable.

Data Availability Statement: Data will be made available upon request from the authors.

Conflicts of Interest: The authors declare no conflict of interest.

References

1. Wurman, P.R.; D'Andrea, R.; Mountz, M. Coordinating hundreds of cooperative, autonomous vehicles in warehouses. *AI Mag.* **2008**, *29*, 9.
2. D'Andrea, R. Guest editorial: A revolution in the warehouse: A retrospective on kiva systems and the grand challenges ahead. *IEEE Trans. Autom. Sci. Eng.* **2012**, *9*, 638–639. [CrossRef]
3. Gadd, M.; Newman, P. A framework for infrastructure-free warehouse navigation. In Proceedings of the 2015 IEEE International Conference on Robotics and Automation (ICRA), Seattle, WA, USA, 26–30 May 2015; IEEE: Piscataway, NJ, USA, 2015; pp. 3271–3278.
4. Wu, C.H.; Tsang, Y.P.; Lee, C.K.M.; Ching, W.K. A Blockchain-IoT Platform for the Smart Pallet Pooling Management. *Sensors* **2021**, *21*, 6310. [CrossRef] [PubMed]
5. Motroni, A.; Buffi, A.; Nepa, P.; Pesi, M.; Congi, A. An Action Classification Method for Forklift Monitoring in Industry 4.0 Scenarios. *Sensors* **2021**, *21*, 5183. [CrossRef] [PubMed]
6. Lamooki, S.R.; Cavuoto, L.A.; Kang, J. Adjustments in Shoulder and Back Kinematics during Repetitive Palletizing Tasks. *Sensors* **2022**, *22*, 5655. [CrossRef] [PubMed]

7. Zhu, X.; Lyu, S.; Wang, X.; Zhao, Q. TPH-YOLOv5: Improved YOLOv5 based on transformer prediction head for object detection on drone-captured scenarios. In Proceedings of the IEEE/CVF International Conference on Computer Vision, Virtual, 11–17 October 2021; pp. 2778–2788.

8. Liebelt, J.; Schmid, C.; Schertler, K. Viewpoint-independent object class detection using 3d feature maps. In Proceedings of the 2008 IEEE Conference on Computer Vision and Pattern Recognition, Anchorage, AK, USA, 23–28 June 2008; IEEE: Piscataway, NJ, USA, 2008; pp. 1–8.

9. Hussain, M.; Al-Aqrabi, H.; Munawar, M.; Hill, R.; Alsboui, T. Domain Feature Mapping with YOLOv7 for Automated Edge-Based Pallet Racking Inspections. *Sensors* **2022**, *22*, 6927. [CrossRef] [PubMed]

10. Creţu-Sîrcu, A.L.; Schiøler, H.; Cederholm, J.P.; Sîrcu, I.; Schjørring, A.; Larrad, I.R.; Berardinelli, G.; Madsen, O. Evaluation and Comparison of Ultrasonic and UWB Technology for Indoor Localization in an Industrial Environment. *Sensors* **2022**, *22*, 2927. [CrossRef] [PubMed]

11. Li, Y.Y.; Chen, X.H.; Ding, G.Y.; Wang, S.; Xu, W.C.; Sun, B.B.; Song, Q. Pallet detection and localization with RGB image and depth data using deep learning techniques. In Proceedings of the 2021 6th International Conference on Automation, Control and Robotics Engineering (CACRE),Dalian, China, 15–17 July 2021; IEEE: Piscataway, NJ, USA, 2021; pp. 306–310.

12. Joo, K.J.; Pyo, J.W.; Ghosh, A.; In, G.G.; Kuc, T.Y. A pallet recognition and rotation algorithm for autonomous logistics vehicle system of a distribution center. In Proceedings of the 2021 21st International Conference on Control, Automation and Systems (ICCAS), Jeju, Korea, 12–15 October 2021; IEEE: Piscataway, NJ, USA, 2021; pp. 1387–1390.

13. Borstell, H.; Kluth, J.; Jaeschke, M.; Plate, C.; Gebert, B.; Richter, K. Pallet monitoring system based on a heterogeneous sensor network for transparent warehouse processes. In Proceedings of the 2014 Sensor Data Fusion: Trends, Solutions, Applications (SDF), Bonn, Germany, 8–10 October 2014; IEEE: Piscataway, NJ, USA, 2014; pp. 1–6.

14. Garibott, G.; Masciangelo, S.; Ilic, M.; Bassino, P. Robolift: A vision guided autonomous fork-lift for pallet handling. In Proceedings of the IEEE/RSJ International Conference on Intelligent Robots and Systems, IROS'96, Osaka, Japan, 8 November 1996; IEEE: Piscataway, NJ, USA, 1996; Volume 2, pp. 656–663.

15. Garibotto, G.; Masciangelo, S.; Bassino, P.; Ilic, M. Computer vision control of an intelligent forklift truck. In Proceedings of the Conference on Intelligent Transportation Systems, Boston, MA, USA, 12 November 1997; IEEE: Piscataway, NJ, USA, 1997; pp. 589–594.

16. Minav, T.A.; Laurila, L.I.; Pyrhönen, J.J. Self-Tuning-Parameter Fuzzy PID Speed Controller Performance in an Electro-Hydraulic Forklift with Different Rule Sets. 2010. Available online: https://www.researchgate.net/profile/Tatiana-Minav/publication/260952145_Self-Tuning-Parameter_Fuzzy_PID_Speed_Controller_Performance_in_an_Electro-Hydraulic_Forklift_with_Different_Rule_Sets/links/5540a17a0cf2736761c280c2/Self-Tuning-Parameter-Fuzzy-PID-Speed-Controller-Performance-in-an-Electro-Hydraulic-Forklift-with-Different-Rule-Sets.pdf (accessed on 9 October 2022).

17. Jiang, Z.; Xiao, B. Electric power steering system control strategy based on robust H control for electric forklift. *Math. Probl. Eng.* **2018**, *2018*, 7614304 . [CrossRef]

18. Wang, C.Y.; Liao, H.Y.M.; Wu, Y.H.; Chen, P.Y.; Hsieh, J.W.; Yeh, I.H. CSPNet: A new backbone that can enhance learning capability of CNN. In Proceedings of the IEEE/CVF Conference on Computer Vision and Pattern Recognition Workshops, Seattle, WA, USA, 14–19 June 2020; pp. 390–391.

19. Redmon, J.; Farhadi, A. Yolov3: An incremental improvement. *arXiv* **2018**, arXiv:1804.02767.

20. Lin, T.Y.; Dollár, P.; Girshick, R.; He, K.; Hariharan, B.; Belongie, S. Feature pyramid networks for object detection. In Proceedings of the IEEE Conference on Computer Vision and Pattern Recognition, Honolulu, OH, USA, 21–26 July 2017; pp. 2117–2125.

21. Woo, S.; Park, J.; Lee, J.Y.; Kweon, I.S. Cbam: Convolutional block attention module. In Proceedings of the European Conference on Computer Vision (ECCV), Munich, Germany, 8–14 September 2018; pp. 3–19.

22. Giesbrecht, J.; Mackay, D.; Collier, J.; Verret, S. *Path Tracking for Unmanned Ground Vehicle Navigation: Implementation and Adaptation of the Pure Pursuit Algorithm*; Technical report; Defence Research and Development Suffield (Alberta): Suffield, AB, Canada, 2005 .

23. Comaniciu, D.; Meer, P. Mean shift: A robust approach toward feature space analysis. *IEEE Trans. Pattern Anal. Mach. Intell.* **2002**, *24*, 603–619. [CrossRef]

24. Lin, T.Y.; Goyal, P.; Girshick, R.; He, K.; Dollár, P. Focal loss for dense object detection. In Proceedings of the IEEE International Conference on Computer Vision, Venice, Italy, 22–29 October 2017; pp. 2980–2988.

25. Duan, K.; Bai, S.; Xie, L.; Qi, H.; Huang, Q.; Tian, Q. Centernet: Keypoint triplets for object detection. In Proceedings of the IEEE/CVF International Conference on Computer Vision, Seoul, Korea, 27 October–2 November 2019; pp. 6569–6578.

26. Zhao, H.; Shi, J.; Qi, X.; Wang, X.; Jia, J. Pyramid scene parsing network. In Proceedings of the IEEE Conference on Computer Vision and Pattern Recognition, Honolulu, OH, USA, 21–26 July 2017; pp. 2881–2890.

27. Qi, C.R.; Yi, L.; Su, H.; Guibas, L.J. Pointnet++: Deep hierarchical feature learning on point sets in a metric space. *arXiv* **2017**, arXiv:1706.02413.

28. Wang, C.; Xu, D.; Zhu, Y.; Martín-Martín, R.; Lu, C.; Fei-Fei, L.; Savarese, S. Densefusion: 6d object pose estimation by iterative dense fusion. In Proceedings of the IEEE/CVF Conference on Computer Vision and Pattern Recognition, Long Beach, CA, USA, 15–20 June 2019; pp. 3343–3352.

29. Arun, K.S.; Huang, T.S.; Blostein, S.D. Least-squares fitting of two 3-D point sets. *IEEE Trans. Pattern Anal. Mach. Intell.* **1987**, *PAMI-9*, 698–700. [CrossRef] [PubMed]

30. Mitchell, W.C.; Staniforth, A.; Scott, I. *Analysis of Ackermann Steering Geometry*; Technical report, SAE Technical Paper; SAE International: Warrendale, PA, USA, 2006.
31. Ziegler, J.; Bender, P.; Dang, T.; Stiller, C. Trajectory planning for Bertha—A local, continuous method. In Proceedings of the 2014 IEEE Intelligent Vehicles Symposium Proceedings, Dearborn, MI, USA, 8–11 June 2014; IEEE: Piscataway, NJ, USA, 2014; pp. 450–457.
32. Xiao, J.; Lu, H.; Zhang, L.; Zhang, J. Pallet recognition and localization using an rgb-d camera. *Int. J. Adv. Robot. Syst.* **2017**, *14*, 1729881417737799. [CrossRef]
33. Baglivo, L.; Biasi, N.; Biral, F.; Bellomo, N.; Bertolazzi, E.; Da Lio, M.; De Cecco, M. Autonomous pallet localization and picking for industrial forklifts: A robust range and look method. *Meas. Sci. Technol.* **2011**, *22*, 085502. [CrossRef]

MDPI

St. Alban-Anlage 66

4052 Basel

Switzerland

Tel. +41 61 683 77 34

Fax +41 61 302 89 18

www.mdpi.com

Sensors Editorial Office

E-mail: sensors@mdpi.com

www.mdpi.com/journal/sensors